WITHDRAWN

Uniformizing Dessins and Belyĭ
Maps via Circle Packing

WITHDRAWN

of the
American Mathematical Society

Number 805

Uniformizing Dessins and Belyĭ Maps via Circle Packing

Philip L. Bowers
Kenneth Stephenson

July 2004 • Volume 170 • Number 805 (second of 4 numbers) • ISSN 0065-9266

American Mathematical Society
Providence, Rhode Island

2000 *Mathematics Subject Classification.*
Primary 52C26, 30F10; Secondary 30C62.

Library of Congress Cataloging-in-Publication Data
Bowers, Philip L., 1956–
Uniformizing dessins and Belyi maps via circle packing/Philip L. Bowers, Kenneth Stephenson.
  p. cm. — (Memoirs of the American Mathematical Society, ISSN 0065-9266 ; no. 805)
"Volume 170, number 805 (second of 4 numbers)."
Includes bibliographical references.
ISBN 0-8218-3523-8 (alk. paper)
  1. Dessins d'enfants (Mathematics). 2. Circle packing. 3. Riemann surfaces. I. Stephenson, Kenneth, 1945– II. Title. III. Series.
QA3.A57 no. 805
[QA613.2]
510 s–dc22
[516′.11]
                                                        2004046104

Memoirs of the American Mathematical Society

This journal is devoted entirely to research in pure and applied mathematics.

Subscription information. The 2004 subscription begins with volume 167 and consists of six mailings, each containing one or more numbers. Subscription prices for 2004 are $583 list, $466 institutional member. A late charge of 10% of the subscription price will be imposed on orders received from nonmembers after January 1 of the subscription year. Subscribers outside the United States and India must pay a postage surcharge of $31; subscribers in India must pay a postage surcharge of $43. Expedited delivery to destinations in North America $35; elsewhere $130. Each number may be ordered separately; *please specify number* when ordering an individual number. For prices and titles of recently released numbers, see the New Publications sections of the *Notices of the American Mathematical Society*.

Back number information. For back issues see the *AMS Catalog of Publications*.

Subscriptions and orders should be addressed to the American Mathematical Society, P. O. Box 845904, Boston, MA 02284-5904, USA. *All orders must be accompanied by payment.* Other correspondence should be addressed to 201 Charles Street, Providence, RI 02904-2294, USA.

Copying and reprinting. Individual readers of this publication, and nonprofit libraries acting for them, are permitted to make fair use of the material, such as to copy a chapter for use in teaching or research. Permission is granted to quote brief passages from this publication in reviews, provided the customary acknowledgment of the source is given.

Republication, systematic copying, or multiple reproduction of any material in this publication is permitted only under license from the American Mathematical Society. Requests for such permission should be addressed to the Acquisitions Department, American Mathematical Society, 201 Charles Street, Providence, Rhode Island 02904-2294, USA. Requests can also be made by e-mail to reprint-permission@ams.org.

Memoirs of the American Mathematical Society is published bimonthly (each volume consisting usually of more than one number) by the American Mathematical Society at 201 Charles Street, Providence, RI 02904-2294, USA. Periodicals postage paid at Providence, RI. Postmaster: Send address changes to Memoirs, American Mathematical Society, 201 Charles Street, Providence, RI 02904-2294, USA.

© 2004 by the American Mathematical Society. All rights reserved.
This publication is indexed in *Science Citation Index*®, *SciSearch*®, *Research Alert*®, *CompuMath Citation Index*®, *Current Contents*®/*Physical, Chemical & Earth Sciences*.
Printed in the United States of America.

∞ The paper used in this book is acid-free and falls within the guidelines established to ensure permanence and durability.
Visit the AMS home page at http://www.ams.org/

10 9 8 7 6 5 4 3 2 1   09 08 07 06 05 04

Contents

List of Tables   ix

List of Figures   xi

Chapter 1.   Introduction   1

Chapter 2.   Dessins d'Enfants   7
    2.1.   Dessins d'enfants   7
    2.2.   The Equilateral Structure   8
    2.3.   The Belyĭ Map   10

Chapter 3.   Discrete Dessins *via* Circle Packing   13
    3.1.   Circle Packings   13
    3.2.   Discrete Dessins   15
    3.3.   Hexagonal Refinement   17
    3.4.   Geometric Lemmas   18

Chapter 4.   Uniformizing Dessins   23
    4.1.   Reflective Structures and Conformal Subdivisions   23
    4.2.   Uniformizing Equilateral Surfaces   27
    4.3.   Convergence of the Belyĭ Maps   33

Chapter 5.   A Menagerie of Dessins d'Enfants   37
    5.1.   Genus 0   37
    5.2.   Genus 1   40
    5.3.   Genus 2   42
    5.4.   Higher Genera   47

Chapter 6.   Computational Issues   55
    6.1.   Dessin Modifications   58

Chapter 7.   Additional Constructions   61
    7.1.   Conformal Tilings   61
    7.2.   The j function   69
    7.3.   Schwarz Triangles   70
    7.4.   Graph Embedding   75

Chapter 8.   Non-equilateral Triangulations   77
    8.1.   Welding Approach   77
    8.2.   Inversive Distance Packings   78

Chapter 9.   The Discrete Option   83

9.1.	Dessins	83
9.2.	Function Theory	85

Chapter 10.	Appendix: Implementation	89
10.1.	A Quick Experiment	89
10.2.	The algorithm	90
10.3.	Accuracy	91

Bibliography	95

Abstract

Grothendieck's theory of *Dessins d'Enfants* involves combinatorially determined affine, reflective, and conformal structures on compact surfaces. In this paper the authors establish the first general method for uniformizing these dessin surfaces and for approximating their associated Belyĭ meromorphic functions.

The paper begins by developing a *discrete* theory of dessins based on circle packing. This theory is surprisingly faithful, even at its coarsest stages, to the geometry of the *classical* theory, and it displays some new sources of richness; in particular, algrebraic number fields enter the theory in a new way. Furthermore, the discrete dessin structures converge to their classical counterparts under a hexagonal refinement scheme. Since the discrete objects are computable, circle packing provides opportunities both for routine experimentation and for large scale explicit computation, as illustrated by a variety of dessin examples up to genus 4 which are computed and displayed.

The paper goes on to discuss uses of discrete conformal geometry with triangulations arising in other situations, such as conformal tilings and discrete meromorphic functions. It concludes by addressing technical and implementation issues and open mathematical questions that they raise.

1991 *Mathematics Subject Classification*. Primary: 52C26, 30F10; Secondary: 30C62.

Key words and phrases. dessins d'enfants, circle packing, conformal subdivision, hexagonal refinement.

Received by the editor December 15, 1997 and in revised form July 3, 2003. The second author gratefully acknowledges support of the National Science Foundation and the Tennessee Science Alliance.

List of Tables

1	Computation details, genus 0 and 1	92
2	Computation details, higher genus	93
3	Timings for the stage-3 Klein surface.	93

List of Figures

1.1	A simple dessin and associated triangulation.	2
1.2	Equilateral and circle-packed surfaces.	3
1.3	A discrete Belyĭ map.	5
2.1	Power map $z \mapsto z^{3/2}$ for closing up four faces.	10
3.1	A coarse discrete Belyĭ pair.	16
3.2	The hexagonal subdivision operation.	18
3.3	A discrete power map.	21
4.1	Conformal subdivision rules.	25
4.2	The reflective subdivision $\lambda(\delta)$.	26
4.3	h_n restricted to $2N$ hexagonal generations.	30
5.1	Dessin 1 and its stage-4 packing in $\mathbb{P}$.	37
5.2	Stereographic projection of Dessin 2.	38
5.3	Dessin 3: A genus 0 Galois orbit.	39
5.4	Successively finer stages of the right-armed dessin.	39
5.5	Stereographic projection of the head of Dessin 3.	40
5.6	A dessin tree.	41
5.7	Dessins 5 and 6: A genus 1 Galois orbit.	42
5.8	Fundamental domains for genus 1 packings.	43
5.9	The Galois conjugate of Dessin 5.	44
5.10	Dessins 7 and 8: Two dessins of genus 2.	45
5.11	Coarse and stage-3 fundamental domains for Dessin 7.	46
5.12	Dessins 8 and 9.	48
5.13	Klein's Hauptfigur.	50
5.14	Two dessins for the Picard surface.	51
5.15	A standard pair of pants.	52
5.16	A symmetric genus 4 surface.	52
5.17	Details of a more generic genus 4 dessin.	53
6.1	Comparing cusps for stages of Example 7.	57
6.2	Dessin moves generating Dessin 6.	60

7.1	A 'regular' pentagonal tiling of the plane.	62
7.2	Pentagonal subdivision rule of Cannon, Floyd, and Parry.	62
7.3	Initial circle packing embeddings.	63
7.4	Schematics of R^{-1} iterates	66
7.5	Fifth root of the schematic.	67
7.6	Dessin of the regular pentagonal tiling.	68
7.7	The j-function	70
7.8	Functions for other triangle groups.	71
7.9	Behavior at a simple branch point.	72
7.10	Schwarz triangles t and T.	73
7.11	120 Schwarz triangles tiling the sphere.	74
7.12	The branched packing $\widetilde{\mathcal{P}}$ and one of its 12 branched flowers.	75
8.1	Discrete conformal welding	79
8.2	Inversive distance packings: random overlaps, random separations	80
8.3	Examples of S-packings and refinements.	82
9.1	Dodecahedral subdivision rules.	85
9.2	Stage 3 rectangle under the dodecahedral subdivision rule.	86
9.3	Combinatorial conformal feedback.	87

CHAPTER 1

Introduction

We are concerned in this paper with structures on triangulated surfaces. Our motivation is the theory of *Dessins d'Enfants*, traced to Grothendieck [23], an intriguing blend of algebra, combinatorics, conformal geometry, and complex function theory.

The creation mythology of the topic posits a child innocently "drawing" on a topological surface. Unbeknownst (presumably) to the child, that simple drawing determines an algebraic number field and its Galois group, a conformal stucture on the surface, a class of meromorphic functions, and a group of companion drawings. This becomes a story, then, of various rigid algebraic, analytic, and geometric structures, inextricably intertwined, but all flowing from simple combinatorics.

The central aim of the study for algrebraists has been a deeper understanding of the absolute Galois group $\mathrm{Gal}(\overline{\mathbb{Q}}/\mathbb{Q})$ and the famous "inverse Galois problem". Of course, any theory bringing together so many topics will invariably inspire other goals, and an extensive literature has developed. We particularly recommend the proceedings [44] for the 1993 Luminy conference for a broad view. Interests are both theoretical and practical: along with the discussion of "braid towers" and Shimura varieties are the efforts of physicists to connect dessins with matrix models in computational studies of Riemann surfaces for string theory.

The authors of the present paper, having worked in circle packing, were drawn to the topic principally through triangulations: combinatorics also lead to rigid geometric structures *via* circle packing. A rich theory has developed around circle packings since their introduction by Thurston in [50] and [51], with particular connections to conformal geometry. One familiar with these developments cannot help but recognize the many parallels with at least the combinatoric and geometric aspects of the theory of dessins. The links promise to enrich the theory and practice of both topics. In addition, there is a visual and numerical side to circle packing which might contribute to the theory of dessins — perhaps introducing an experimental aspect.

To introduce the viewpoint of this paper, it will help to review briefly the elements of the theory of dessins. A *dessin d'enfant* D is basically a finite connected graph on a (compact orientable) surface S. Associated with D is a *canonical triangulation* T of S; this is our fundamental combinatorial data. A simple genus 0 dessin and its triangulation are given in Figure 1.1.

The triangulation T in turn imposes a conformal structure on S making it into a Riemann surface S_D; this is accomplished by constructing a model for S by pasting together euclidean unit-sided equilateral triangles in the pattern encoded in T — a so-called *equilateral surface*. The triangulation is such that alternate faces may be shaded; if the unshaded faces are mapped conformally to the upper

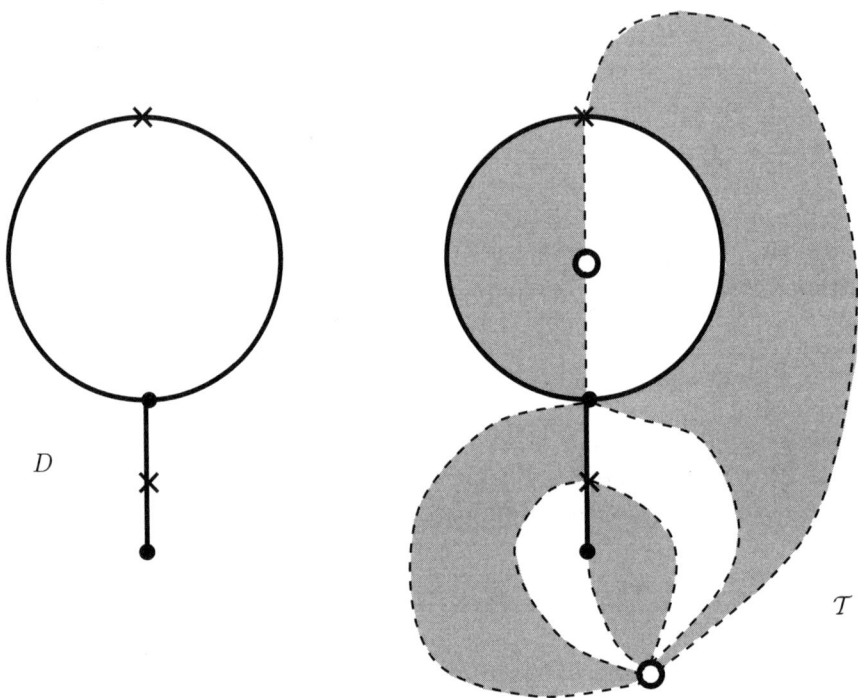

FIGURE 1.1. A simple dessin and associated triangulation.

half-plane, the shaded faces to the lower half-plane, and appropriate identifications are made along the intervals $[\infty, 0], [0, 1], [1, \infty]$, then one obtains a meromorphic function $B_D : S_D \longrightarrow \mathbb{P}$ branching only over $\{0, 1, \infty\}$; here $\mathbb{P}$ is complex projective space, also known as the Riemann sphere. The structures are conveniently packaged in the so-called *Belyĭ pair* (S_D, B_D); indeed, the dessin D itself is in there, being (isomorphic to) the set $B_D^{-1}[0, 1] \subset S_D$. Moreover, the Riemann surface S_D, as an algebraic surface, has a defining equation whose coefficients lie in an algebraic number field F (a finite algebraic extension of the rationals).

For surfaces of positive genus, the results may be summarized in a very striking form. The most difficult of the implications is due to Belyĭ, so this is commonly referred to as

BELYĬ'S THEOREM. *For a Riemann surface R of genus $g \geq 1$, the following statements are equivalent:*

(a) There exists a defining equation for R whose coefficients lie in an algebraic number field F over the rationals.

(b) There exists a nonconstant meromorphic function $f : R \longrightarrow \mathbb{P}$ which branches only over the points $\{0, 1, \infty\}$.

(c) R is conformally equivalent to an equilateral surface.

In other words, we have a distinguished class of Riemann surfaces which simultaneously enjoys algebraic, function-theoretic, and combinatoric characterizations. Equivalent to all these, $R = S_D$ for some dessin D. Thus, all these consequences flow from a simple drawing.

Let us now turn to circle packing and what we will refer to as the "discrete" setting. A *circle packing* is a configuration of circles realizing a specified pattern of tangencies. It enjoys dual combinatoric and geometric natures: the "pattern" of tangencies is encoded as an abstract triangulation of a surface, while circle radii provide the geometry. We paraphrase the central theoretical pivot, with terminology to be explained later.

THEOREM. *Let T be a simplicial triangulation of a compact oriented topological surface S. Then there exists a unique Riemann surface homeomorphic to S that supports a univalent circle packing P with the combinatorics of T.*

Thus circle packings provide an alternate way in which abstract combinatorics determine a rigid conformal structure.

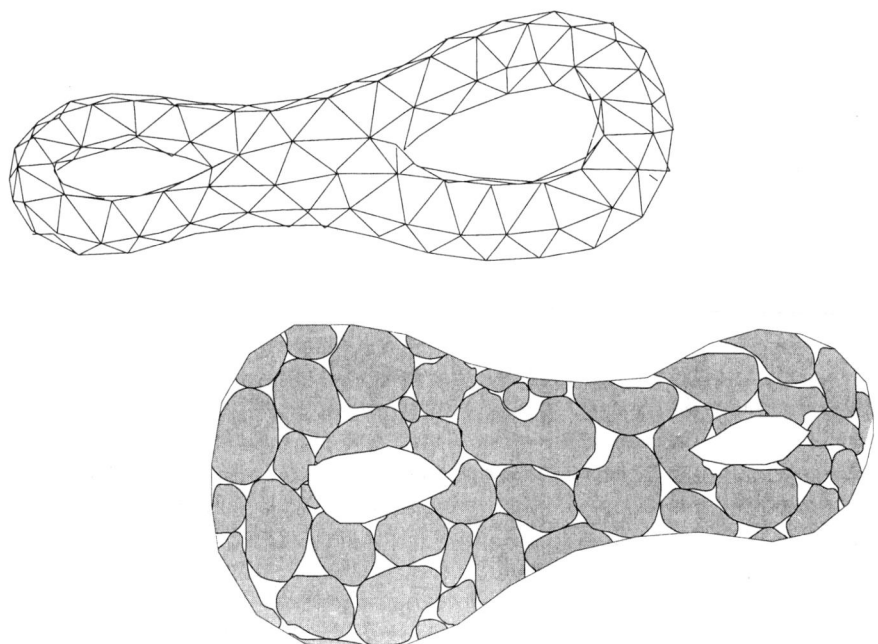

FIGURE 1.2. Equilateral and circle-packed surfaces.

Let's now begin again with a dessin D on a surface S. Generate the associated triangulation T as before (though for technical reasons we work instead with the barycentric subdivision of T). The Circle Packing Theorem provides a conformal structure for S, giving a Riemann surface, denoted s_D, which supports a circle packing P for T. Again, we may shade alternate faces of T, defined geometrically by the circles. Circles corresponding to the unshaded faces can be identified with circles in the sphere which pack the upper half-plane, those corresponding to the shaded faces with circles which pack the lower half-plane, and the edge circles can be appropriately identified along the intervals $[0, 1], [1, \infty],$ and $[\infty, 0]$ to form a circle packing $\mathcal{Q}$ of $\mathbb{P}$. The identification of $\mathcal{P}$ with $\mathcal{Q}$ yields a mapping $b_D : s_D \longrightarrow \mathbb{P}$ which acts as a discrete meromorphic function branching only over $0, 1, \infty$. Thus we arrive at a discrete Belyĭ pair (s_D, b_D) associated with our dessin.

In summary, then, a dessin can be associated with rigid conformal data in TWO parallel ways, *via* classical Belyĭ pairs or *via* their discrete analogues.

First Objective: *to develop the discrete theory, emphasizing these parallels in combinatorics and geometry.*

However, our discrete objects not only mimic their classical counterparts, but also approximate them. We prove that certain refinements $\mathcal{T}_n$ of a triangulation $\mathcal{T}$, while inducing the identical *classical* Belyĭ pair (S_D, B_D), will lead to new *discrete* Belyĭ pairs $(s_D^{(n)}, b_D^{(n)})$ based on "finer" circle packings. We prove that under successive refinement,
$$(s_D^{(n)}, b_D^{(n)}) \longrightarrow (S_D, B_D), \text{ as } n \to \infty.$$
In other words, the Riemann surfaces $s_D^{(n)}$ converge to S_D in Teichmüller space, while the discrete Belyĭ maps $b_D^{(n)}$ converge uniformly on compacta to B_D. The discrete objects have the advantage that they are effectively computable.

Second Objective: *to prove that the objects of the discrete theory generated by circle packings uniformize classical dessin surfaces and approximate classical Belyĭ maps.*

It should be noted that there are varying uses of the term "uniformize" in the dessin literature: For a given dessin surface S_D, this often refers to the structure it inherits from a classical triangle group by modding out a covering group; this is actually a conformal structure on the punctured surface $S_D \backslash V$, where V is the set of vertices of $\mathcal{T}$. That is not how we use the term; we uniformize the full surface S_D in the classical function theory sense. Despite very concrete descriptions of these surfaces, it has been essentially impossible to compute such uniformizations in the past, aside from certain highly symmetric situations.

We might exploit the visual nature of circle packings here with an early example — details will come later. Let's consider the dessin of Figure 1.1. Figure 1.3(a) illustrates the circle packing of $\mathbb{P}$ associated with three stages of refinement of $\mathcal{T}$; the edges of $\mathcal{T}$ are drawn in, with the heavy edges being those of the dessin D, and appropriate faces are shaded. Figure 1.3(b) illustrates the packings of the upper/lower half planes at the corresponding refinement stage. The discrete meromorphic function $b_D^{(3)}$ identifies each circle on the left with a corresponding circle on the right. (This defines a 4-sheeted covering of $\mathbb{P}$, so each circle on the right is the image of four circles on the left.) The discrete dessin is nearly conformally correct, and in positive genus examples the surface will be nearly equal in modulus to the classical surface. This illustrates our

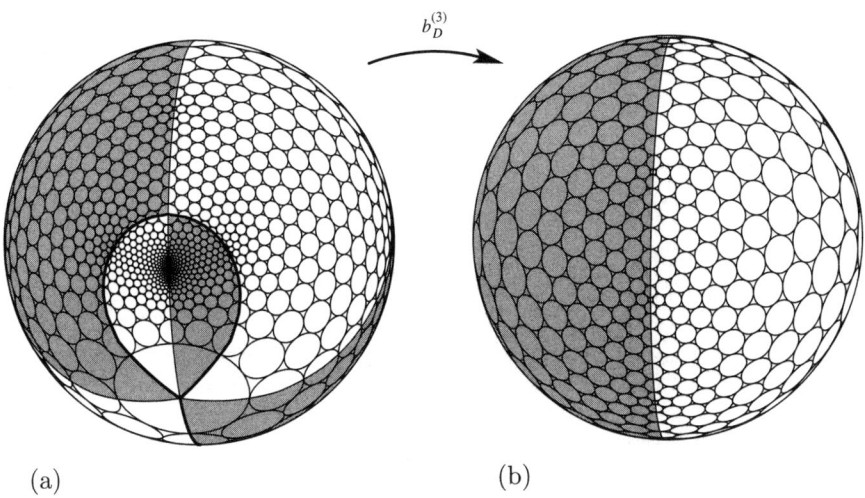

Figure 1.3. A discrete Belyĭ map.

Third Objective: *to provide a variety of examples and discuss the computational issues associated with the theory.*

A notable loss in the discrete setting is the direct connection to number fields. Recall that Belyĭ's Theorem provides an elegant algebraic characterization of the countable dense set of "equilateral" points of Teichmüller space. There is as yet no characterization of the "circle packing" points, also countable and dense; this characterization appears to be a challenging issue. The circle packing points are, of course, *indirectly* associated with number fields *via* the equilateral surfaces. However, it also turns out that algebraic numbers enter directly, but in a new way — as entries in the covering group in $PSL(2,\mathbb{C})$. There are a number of open questions here, and we will return to this issue in Chapter 9.

Triangulations arise in a variety of conformal settings not directly related to dessins and many of the techniques we develop can be modified and extended. We will look at discretizing the classical j-function, at building non-Belyĭ meromorphic functions, and at a conformal tiling construction of Cannon, Floyd, Parry, and Kenyon combining a Belyĭ-type function with rational iteration. This contributes to our

Fourth Objective: *to propagandize for the "discrete" option — that is, to suggest an appreciation of the discrete theory for its own interest, variety, and intrinsic beauty.*

There are many new avenues open for investigation, and it would be a shame to hobble the theory by tethering it too rigidly to its classical roots.

Now for a brief outline of our monograph. Chapter 2 is concerned with dessins in the classical setting (we use the term "classical" only to distinguish it from the discrete setting), and we review the basic terminology, notation, and theory used in the sequel. In Chapter 3 we develop the parallels in the discrete setting, beginning with the basic theory of circle packing, defining discrete Belyĭ pairs, and ending with the geometry of hex refinement.

Chapter 4 contains the monograph's main theoretical results. We introduce the key notions of *reflective structures* and their *conformal subdivisions* for triangulated surfaces. This permits refinement of a triangulation without changing its associated conformal structure. We then prove (Theorems 4.1 and 4.2) that the discrete Belyĭ pairs for successive refinements of a dessin triangulation converge to the classical Belyĭ pair for the dessin. In other words, using circle packings we simultaneously uniformize the associated Riemann surface and approximate its Belyĭ map.

Chapter 5 is a Menagerie of examples of dessins, organized by genus. We encourage the reader to make an early visit. We illustrate several examples from the dessin literature, classical surfaces such as Klein's and Picard's, and new examples up to genus 4. These show the type and quality of information available in this discrete approach. We go more deeply into computational issues in Chapter 6, addressing the accuracy and faithfulness of the discrete objects to their classical counterparts (though nitty-gritty details are placed in the Appendix). We define a "dessin modification" procedure which, coupled with speedy computation and display, may provide new insights into the behavior of dessins.

Chapters 7 and 8 extend our methods to other settings in which triangulations induce conformal structures and to nonequilateral triangulations. We construct conformal tilings, the classical j-function, and illustrate the goal for more general meromorphic functions using branched tilings of the sphere by Schwarz triangles. For handling nonequilateral triangulations, we describe a welding method and speculate on the use of "inversive distance" circle packings.

The concluding Chapter 9 summarizes the internals of discrete geometry in the circle packing setting, its faithfulness to its classical counterparts, its experimental strengths, and its intrinsic interest and beauty. We highlight several questions about circle packing: the characterization of packable Riemann surfaces, the surprising accuracy of even coarse circle packing structures, and new connections to algebraic numbers arising with discrete dessins, to name a few. We close with a "conformal feedback" mechanism inspired by experiments with conformal tilings which illustrates that discrete conformal geometry brings new ideas and techniques to the table. It is our way to encourage tourism by suggesting that the world of discrete conformal geometry has attractions beyond the approximation of the classical.

The authors express their thanks to Jack Quine, Robert Varley, Chuck Collins, and David Singerman for helpful conversations and to Alan Beardon, Keith Carne, and the University of Cambridge for their hospitality during the authors' sabbaticals when this work began. (For readers wanting more computational details, we have included an Appendix discussing alogrithms, complexity, visualization, and so forth. The illustrations in the paper were generated using packages `CirclePack` and `DesPack` available from the second author.)

CHAPTER 2

Dessins d'Enfants

Our work involves various procedures for constructing Riemann surfaces (with extra structure) from purely combinatorial data. In this section we review the fundamentals of Grothendieck's theory of *dessins d'enfants* which motivate the developments of the paper. Dessin theory is of recent origin and we refer to it as "classical" only to distinguish it from the coming "discrete" theory.

Authors differ slightly in the terminology they employ for dessins d'enfants. We adopt that of [44, p.4] and refer the reader to [44] for more extensive treatments of the topic and, in particular, for a discussion of the deep and intimate connections between dessins d'enfants, Belyĭ maps, surfaces defined over number fields, and the absolute Galois group $\mathrm{Gal}(\overline{\mathbb{Q}}/\mathbb{Q})$. Here we recall terminology, fix notation, and state results which set the pattern for the remainder of the paper.

2.1. Dessins d'enfants

DEFINITION. *A **dessin d'enfant**, or **dessin** for short, is an oriented, closed (compact and connected with empty boundary) topological surface S equipped with a finite embedded graph D for which*

(1) *the graph D is connected,*
(2) *the complement $S \backslash D$ is a (finite) collection of open 2-cells,*
(3) *the vertices of D are 2-colorable; i.e., the vertex set V decomposes as the disjoint union of two nonempty collections, V_0 and V_1, such that every edge of D has one vertex in V_0 and the other in V_1.*

Requirement (3) implies that D has no loops, though there may be multiple edges. The vertices in V_i are called the i-**vertices**, for $i = 0, 1$. In diagrams of dessins, we use a solid dot • to mark 0-vertices and a small × to mark 1-vertices. We consider dessins (S_1, D_1) and (S_2, D_2) to be **equivalent** if there exists an orientation preserving homeomorphism of S_1 to S_2 whose restriction to D_1 is a color-preserving graph isomorphism of D_1 to D_2. We make no distinction between equivalent dessins.

A dessin (S, D), or just D if S is understood from the context, is **pre-clean** if each 1-vertex has valence at most 2 and is **clean** if each 1-vertex has valence equal to 2. There is a standard way to get a clean dessin from a graph G — what Grothendieck described as a "drawing". For simplicity, nearly all our examples will be of this type, although the methods apply to all dessins, and even more general situations. If G is embedded in a surface S and (1) and (2) hold, one obtains a clean dessin D by adding a × to each edge, so V_0 is the set of original vertices of G and V_1 the set of added vertices. There is no loop-free restriction on the graph G.

By **triangulation** of a surface S we mean a regular cellular decomposition $\mathcal{T}$ of S with each 2-cell a topological triangle so that two 2-cells, when they meet, meet in a subcomplex of the 1-skeleton $\mathcal{T}^{(1)}$. If it happens that two faces can meet only at a single vertex or along a single edge, we refer to $\mathcal{T}$ as a **simplicial triangulation** of S. In general, however, a triangulation might have faces meeting along the union of two sides, along one common side and a common opposite vertex, along three common vertices but no sides, or even along all three sides. Notice that we view a triangulation as a structure *in situ*, as a collection of actual subsets of S — vertices consisting of points of S, edges consisting of arcs in S, and faces consisting of 2-cells in S. Abstract triangulations — purely combinatorial objects — will always have surfaces as geometric realizations. The **degree** of a triangulation is the maximum of the degrees of its vertices; that is, the largest number of edges emanating from any one vertex.

Associated with a dessin (S, D) is a **canonical triangulation** of S, denoted $\mathcal{T} = \mathcal{T}(D)$, described as follows. The vertex set of $\mathcal{T}$ is the disjoint union $V_0 \cup V_1 \cup V_\infty$, where V_0 and V_1 are as before and V_∞ consists of a collection of points from $S \backslash D$, one point from each (2-cell) component of $S \backslash D$. The edges of $\mathcal{T}$ consist of the edges of D along with edges formed by connecting vertices in $V_0 \cup V_1$ with those of V_∞ as follows. For $v \in V_0 \cup V_1$, let Δ be a disc neighborhood of v that meets D only at points of $\text{st}(v, D)$; i.e., at points in the edges of D incident with v. The edges of D divide Δ into d open sectors, where $d = \deg(v, D)$. Add, for each sector Σ, an edge emanating from v, meeting D only at v, and traveling along an arc through Σ into the 2-cell component of $S \backslash D$ containing Σ until it meets the vertex in V_∞ in that component. We proceed, of course, so that new edges, if they meet, meet only at the vertices in V_∞. For the general dessin on the left in Figure 1.1, the procedure produces the cellular decomposition of $S = \mathbb{P}$ on the right. As we shall do subsequently, we have marked the ∞-**vertices**, the elements of V_∞, with open dots ∘. The open faces of $\mathcal{T}$ consist of the components of $X \backslash \overline{E}$, where $\overline{E}$ denotes the union of the edges of $\mathcal{T}$. Notice that each face of $\mathcal{T}$ is a topological triangle with three vertices, one of each type, and a circuit of three edges, one of type ●———×, one of type ●———∘, and another of type ×———∘. The faces are naturally partitioned into two collections by the orientation on S, depending on whether or not the circuit ∘———●———×———∘ around the boundary of a face is compatible with the orientation the face inherits from the orientation of S. If the circuit ∘———●———×———∘ is compatible, we call the face a $(+)$triangle; otherwise a $(-)$triangle. In diagrams, we shall shade in the $(-)$triangles, as in Figure 1.1. Notice that each vertex of $\mathcal{T}$ has even degree, and no two $(+)$triangles (resp. $(-)$triangles) share a common edge. Any two canonical triangulations of (S, D) are not only combinatorially equivalent, but isotopic, so we may speak of **the** canonical triangulation of (S, D).

2.2. The Equilateral Structure

Up to this point, a dessin represents purely combinatorial data. However, it can be realized geometrically as a piecewise equilateral surface.

DEFINITION. *Let $\mathcal{T}$ be a triangulation of the topological surface S (possibly with boundary). An **equilateral metric structured on** $\mathcal{T}$ is a piecewise euclidean distance function on S in which each edge of $\mathcal{T}$ is isometric with the unit interval $[0, 1]$ and each face of $\mathcal{T}$ is isometric with a euclidean unit equilateral triangle. The surface S equipped with such a metric is termed an **equilateral surface structured on** $\mathcal{T}$, and is denoted as $|\mathcal{T}|_{eq}$.*

2.2. THE EQUILATERAL STRUCTURE

One can always construct such a metric, defining it first on the 1-skeleton, extending locally so that each face becomes a euclidean unit equilateral triangle, and finally defining the distance between points as the length of a shortest path between them, which always exists. We could equally well use edges of length $\varepsilon > 0$, in which case we denote the surface by $|\mathcal{T}|_{eq}^{\varepsilon}$; this simply respresents a scaling of the metric by ε.

One next defines a compatible conformal structure. Each edge $e \in \mathcal{T}$ is shared by two faces $f_{\pm}$; let $U_e \subset S$ denote the open set $f_+^{\circ} \cup e^{\circ} \cup f_-^{\circ}$. Define $\Omega \subset \mathbb{C}$ to be the interior of the union $\triangle \cup \overline{\triangle}$, where $\triangle$ is a unit equilateral triangle with vertices $\langle 0, 1, \frac{(1+\sqrt{3}i)}{2}\rangle$. From the definition of our equilateral metric, it is evident that there exists an orientation-preserving, isometric map φ_e of U_e onto Ω. The charts $\{(U_e, \varphi_e) : e \in \mathcal{T}\}$ provide $S \backslash V$ (where V is the vertex set of $\mathcal{T}$) with a covering by compatible analytic charts determining a complex atlas. It remains to define an analytic chart in the neighborhood of each vertex.

Given $v \in V$, let U_v be the open metric disc of radius $1/2$ centered at v. If v has degree k, then there is a closed chain $f_1, \cdots, f_k$ of faces about v, with $f_{k+1} = f_1$; U_v is a topological disc lying in the union of these faces and containing no vertices other than v. Identify f_1 isometrically with a unit equilateral triangle Δ_1 with vertices $\langle 0, 1, \frac{(1+\sqrt{3}i)}{2}\rangle$, 0 being associated with v. Successively identify $f_2, \cdots, f_k$ isometrically with unit equilateral triangles $\Delta_2, \cdots, \Delta_k$, where each Δ_j shares an edge with Δ_{j-1}. The resulting map ψ_v is well-defined on $\bigcup_{j=1}^{k} f_j$, except possibly on the edge e shared by f_1 and f_k. When $k = 6$, this construction closes up at e, $\phi_v = \psi_v|_{U_v}$ is an isometry between U_v and $\Omega = \{z : |z| < 1/2\}$, and we have our chart (U_v, ϕ_v). In general, however, we follow ψ_v with a power map, namely $g : z \mapsto z^p$ where $p = 6/k$, which precisely closes up the chain of images of $f_1, \cdots, f_k$. The ambiguity in g when $k \neq 6$ is resolved by specifying the principal branch of z^p on Δ_1 and then extending analytically to $\Delta_2, \cdots, \Delta_k$ in succession. The power p is so chosen that $\phi_v = g \circ \psi_v$ is well defined on the edge e and maps U_v one-to-one onto $\Omega = \{z : |z| < (1/2)^p\}$. Since g is conformal away from $0 = \psi_v(v)$ and ψ_v is locally isometric away from v, ϕ_v is compatible in $U_v \backslash \{v\}$ with the charts already defined on $S \backslash V$. That is, (U_v, ϕ_v) may be taken as our chart at v. Sets U_v are disjoint for distinct vertices v, so adding such a chart at each vertex gives us an atlas and hence a conformal structure on all of S. (See, e.g., [**3**, §3.3]).

Figure 2.1 illustrates the situation when $k = 4$. On the left are the four equilateral triangles $\Delta_1, \cdots, \Delta_4$; the placement of edge e is not well defined until we apply the power map $z \mapsto z^{3/2}$, which closes up the images. The triangles are ruled to give some feel for the mapping behavior.

DEFINITION. *In general, if $\mathcal{T}$ is a triangulation of a topological surface S, the Riemann surface associated with the equilateral surface structured on $\mathcal{T}$ is called the **equilateral surface** for $\mathcal{T}$ and denoted $S_{\mathcal{T}}$. In case $\mathcal{T} = \mathcal{T}(D)$, the canonical triangulation for dessin D, we refer to this Riemann surface as the **dessin surface** and denote it by S_D.*

Note that $\mathcal{T}$ will be treated as an *in situ* triangulation of $S_{\mathcal{T}}$. As such, it is uniquely determined up to conformal automorphisms and will be treated as fixed; when necessary, a convenient normalization may be imposed. We will be investigating the "reflective" nature of these structures in the sequel.

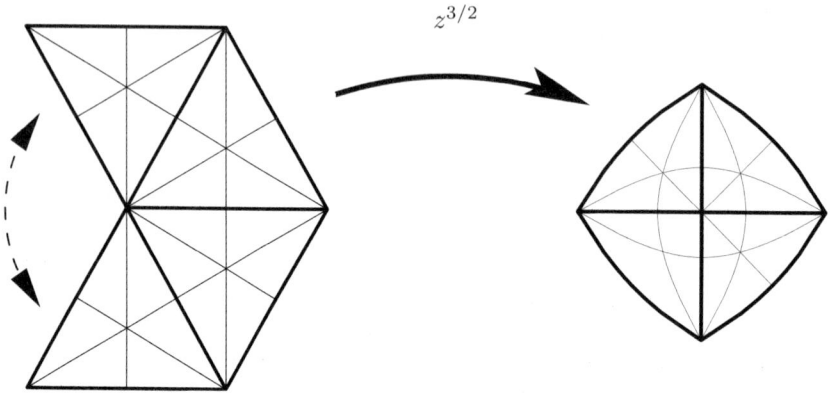

FIGURE 2.1. Power map $z \mapsto z^{3/2}$ for closing up four faces.

2.3. The Belyĭ Map

A **Belyĭ map** of a Riemann surface R is a meromorphic function $B : R \longrightarrow \mathbb{P}$ that is ramified only over 0, 1, and ∞. (In other words, if B fails to be locally one-to-one at a point $z \in R$, then $B(z)$ equals 0, 1, or ∞.)

Using Schwarz reflection, one constructs a canonical Belyĭ map for a given dessin surface S_D as follows. A dessin face $f \in \mathcal{T}(D)$ is a topological triangle with a conformal structure inherited from S_D; consequently there exists a unique conformal mapping of the interior of f onto an open hemisphere of $\mathbb{P}$ that extends continuously to ∂f and carries the j-vertex to $j \in \mathbb{P}$, $j = 0, 1, \infty$. Orientation guarantees that $(+)$triangles are mapped to the upper half-plane, $(-)$triangles to the lower. Define B on the interior of each face f to be the associated conformal map. If faces $f_\pm$ share an edge e of $\mathcal{T}(D)$, recall the map $\varphi_e : U_e \longrightarrow \Omega$ defined above. If $g = B \circ \varphi_e^{-1}$, then the restriction of g to $\triangle \subset \Omega$ (resp. $\overline{\triangle} \subset \Omega$) is a conformal map onto the upper (resp. lower) half-plane and g extends to the interval $[0, 1]$ and is real there. By Schwarz reflection, g extends analytically to a conformal map on all of Ω. In particular, then, B extends analytically across the open edge e and is locally one-to-one. That is, B is a smooth analytic covering map of $S_D \backslash V$ onto the thrice punctured sphere $\mathbb{P} \backslash \{0, 1, \infty\}$. The isolated singularities are removable, with values in $\{0, 1, \infty\}$ so B is a meromorphic function on S_D which can branch only over $\{0, 1, \infty\}$.

DEFINITION. *Given a dessin (S, D), the meromorphic function $B : S_D \longrightarrow \mathbb{P}$ defined as above is termed the **Belyĭ map for** D and will be denoted B_D. Note that B_D is uniquely determined up to conformal automorphisms of S_D. The pair (S_D, B_D) is called a (classical) **Belyĭ pair**. The set $B_D^{-1}[0, 1] \subset S_D$ is D and in the presence of the conformal structure will be called a **conformally correct** dessin.*

Note that any pair (R, B), where R is a compact Riemann surface and $B : R \longrightarrow \mathbb{P}$ is a Belyĭ map, will be a Belyĭ pair for the dessin $D = B^{-1}[0, 1] \subset R$.

In summary, then, the classical theory takes one from elementary and purely combinatorial data to a rigid geometric surface and an explicit associated meromorphic function. Recall that Belyĭ's Theorem further associates with each Belyĭ pair an algebraic number field; we will not make direct use of this.

CHAPTER 3

Discrete Dessins *via* Circle Packing

In this section we develop the discrete parallels of the classical objects in the previous section. First we review the background on circle packings pertinent to our task of uniformization; the basic references for this material are [**4, 6, 20, 25, 38, 43, 49, 50**]. For background on Riemann surfaces, see [**3, 21**].

3.1. Circle Packings

Circle packings are configurations of circles realizing specified patterns of tangency. The key to understanding them lies in recognizing their dual natures, *geometric* in the radii of the circles, but *combinatoric* in their required pattern of contact. We first discuss geometry on Riemann surfaces, then the encoding of the combinatorics.

Recall that every Riemann surface admits an essentially unique complete metric of constant curvature 0 or ±1 that is compatible with its conformal structure in the very strong sense that isothermal local coordinates determined by the metric cover the surface with analytic charts from the complete complex atlas of the surface [**56**]. This metric will be termed an **intrinsic metric**, and one of its most useful properties is that it is **conformal**: that is, if two smooth curves in the surface intersect at a point p and if (U, ϕ) is any chart in the complex atlas with $p \in U$, then the angle between the curves at p as determined by the Riemannian metric on the surface agrees with the angle between their images under ϕ in the plane. The intrinsic metric is unique up to conformal automorphisms of the surface.

Let ρ be an intrinsic metric for the Riemann surface R. A metric disk in R of radius $r > 0$ and center $z \in R$ is a topological 2-cell D in R for which

$$D = \{w \in R : \rho(z, w) \leq r\}.$$

The boundary of D, $\partial D = \{w \in R : \rho(z, w) = r\}$, is then a topological circle and is said to be a metric circle in R with center z. The term **circle** will always refer to a metric circle in the intrinsic metric of the appropriate Riemann surface.

All our circle packings will be associated with triangulations. More formally, suppose $\mathcal{K}$ is an abstract simplicial 2-complex that triangulates an orientable topological surface; we use the abbreviated term **complex**. A configuration $\mathcal{P}$ of circles in a Riemann surface R is a **circle packing for** $\mathcal{K}$ provided

(1) $\mathcal{P}$ contains a circle c_v associated with each vertex $v \in \mathcal{K}^{(0)}$,
(2) c_v is (externally) tangent to c_w at a single point whenever vw is an edge of $\mathcal{K}$, and
(3) the geodesic segments connecting the centers of c_u, c_v, and c_w bound a metric triangle in R whenever uvw is a face of $\mathcal{K}$.

The geometric 2-complex in R determined by connecting centers of tangent circles in $\mathcal{P}$ by geodesic segments is called the **carrier** of $\mathcal{P}$, written $\mathrm{carr}(\mathcal{P})$.

In this paper, $\mathcal{P}$ will always be **univalent**, meaning that the circles have mutually disjoint interiors, and $\mathrm{carr}(\mathcal{P})$ fills R. In this case, $\mathrm{carr}(\mathcal{P})$ provides a concrete realization of the abstract complex $\mathcal{K}$ as a geodesic triangulation of the surface R, each face a metric triangle with geodesic sides. Considerably more striking is the fact that the abstract complex $\mathcal{K}$ actually determines uniquely the Riemann surface R; this is the content of the following basic existence-uniqueness result, a proof of which may be found in [**4, 50**]. We state the result for compact orientable surfaces.

CIRCLE PACKING THEOREM. *Let $\mathcal{K}$ be an abstract simplicial complex triangulating a compact, connected, oriented topological surface S. Then S admits uniquely the structure of a Riemann surface R that supports a univalent circle packing $\mathcal{P}$ for $\mathcal{K}$; moreover, the circle packing $\mathcal{P}$ is unique up to conformal automorphisms of R.*

If $\mathcal{K}$ and S are as in the Circle Packing Theorem, the Riemann surface R will be denoted by $|\mathcal{K}|_{cp}$ and the packing $\mathcal{P}$, with circles measured in the intrinsic metric of $|\mathcal{K}|_{cp}$ (constant curvature $0, \pm 1$), will be denoted $\mathcal{P}_{\mathcal{K}}$. We call $|\mathcal{K}|_{cp}$ the **circle packing surface** determined by $\mathcal{K}$ and refer to its conformal structure as a **discrete** conformal structure; it is a perfectly legitimate conformal structure, the adjective "discrete" simply acknowledges its source. When S has genus 0, then $|\mathcal{K}|_{cp}$ is the Riemann sphere; an additional normalization of $\mathcal{P}_{\mathcal{K}}$ will typically be necessary, since circle radii and centers (and hence $\mathrm{carr}(\mathcal{P}_{\mathcal{K}})$) are not invariant under conformal automorphisms.

Circle packings exist in great variety, and mappings among them have been shown to display quantitative and qualitative properties associated with analytic functions. Indeed, the proof of the Circle Packing Theorem follows closely the lines of classical covering theory and the Riemann Mapping Theorem. We require only parts of this broader theory.

Let's begin with covering theory. A complex $\mathcal{K}$ triangulating a compact orientable surface S can always be "lifted" to a simply connected complex $\widetilde{\mathcal{K}}$ triangulating the universal covering surface of S. The Discrete Uniformization Theorem of [**4**] gives an essentially unique **maximal** univalent circle packing $\widetilde{\mathcal{P}}$ for $\widetilde{\mathcal{K}}$. If $\widetilde{\mathcal{K}}$ is finite, then $\widetilde{\mathcal{K}} = \mathcal{K}$, $\widetilde{\mathcal{P}} = \mathcal{P}$ packs the Riemann sphere, and we are finished. Otherwise, $\widetilde{\mathcal{P}}$ is an infinite packing whose carrier is either the euclidean or the hyperbolic plane. If we denote this plane by $\mathcal{D}$, then $\widetilde{\mathcal{P}}$ is unique up to elements of $\mathrm{Aut}(\mathcal{D})$, the conformal automorphisms (Möbius transformations) of $\mathcal{D}$. The simplicial deck transformations of the abstract covering $\widetilde{\mathcal{K}} \longrightarrow \mathcal{K}$ are identified with a discrete group $\Gamma \subset \mathrm{Aut}(\mathcal{D})$ which leaves $\widetilde{\mathcal{P}}$ invariant. Therefore, $\widetilde{\mathcal{P}}/\Gamma$ may be identified with a circle packing $\mathcal{P}$ for $\mathcal{K}$ in the Riemann surface $R = \mathcal{D}/\Gamma$. Even the radii are unaffected by this identification, since the intrinsic metric of R is precisely that inherited from the metric of $\mathcal{D}$. When we come to examples in our Menagerie, the only practical way to display packings of surfaces is to identify them with fundamental domains within their universal covering packings.

Next, let us consider maps between circle packings, the identification map $\widetilde{\mathcal{P}} \longrightarrow \widetilde{\mathcal{P}}/\Gamma = \mathcal{P}$ being a prime example. In general, let $\mathcal{P}$ and $\mathcal{Q}$ be circle packings lying in (possibly distinct) Riemann surfaces. A map $f : \mathcal{P} \longrightarrow \mathcal{Q}$ which preserves tangencies and orientation is termed a **discrete analytic function**; if $\mathcal{Q}$ lies on the Riemann sphere, f is called a **discrete meromorphic** function, and we will see examples of these shortly. The projection $\pi : \widetilde{\mathcal{P}} \longrightarrow \widetilde{\mathcal{P}}/\Gamma$ is a **discrete covering map**. The map f clearly induces a simplicial map of $\text{carr}(\mathcal{P})$ to $\text{carr}(\mathcal{Q})$. We will abuse notation and use f to denote both the set map $f : \mathcal{P} \longrightarrow \mathcal{Q}$ and the associated point mapping $f : \text{carr}(\mathcal{P}) \longrightarrow \text{carr}(\mathcal{Q})$ defined to carry the center of each circle $c \in \mathcal{P}$ to the center of $f(c) \in \mathcal{Q}$ and then extended via barycentric coordinates. The function f is basically a simplicial map which acquires the geometry imposed by the circle packings of its domain and range.

REMARK. Natural barycentric coordinates for geodesic triangles exist in each of our classical geometries; in the hyperbolic setting, they are most easily described in the hyperboloid model, where geodesics are the intersections of planes through the origin with the hyperboloid sheet (see [**56**, **42**]). In particular, as a point mapping, f is a continuous, orientation preserving, light interior mapping of $\text{carr}(\mathcal{P})$. It maps each edge of $\text{carr}(\mathcal{P})$ "convexly" onto an edge of $\text{carr}(\mathcal{Q})$ in the metric sense. We will see later that lower bounds on the angles of the faces in $\text{carr}(\mathcal{P})$ and $\text{carr}(\mathcal{Q})$ also give bounds on the quasiconformal dilatation of f on the interiors of the faces.

Discrete analytic functions can fail to be locally one-to-one, even between univalent packings. Consider $f : \mathcal{P} \longrightarrow \mathcal{Q}$ with $\mathcal{P}$ and $\mathcal{Q}$ univalent. Let $c_0 \in \mathcal{P}$; the flower of c_0 consists of c_0 as the **center** and the sequence $\{c_1, c_2, \cdots, c_k\}$ of petal circles. The petals are tangent to c_0, successively tangent to one another, and wrap around c_0 once in the positive direction. The image circles $\{f(c_1), f(c_2), \cdots, f(c_k)\}$ in $\mathcal{Q}$ will necessarily be petals of the flower of $f(c_0)$. By orientation they must wrap in the positive direction about $f(c_0)$; if they wrap $n \geq 2$ times around $f(c_0)$, we say that f has a (discrete) **branch point** at c_0 of **order** $n - 1$. For instance, if 12 "petals" around c_0 are mapped to four about $f(c_0)$, the result is a branch point of order 2, with each petal of $f(c_0)$ the image of three petals of c_0.

Those curious about the practical side of circle packing should see the Appendix, which briefly discusses numerical algorithms, implementation, and software. For further connections between circle packing and analytic functions, see [**4**, **20**, **43**] and references therein.

3.2. Discrete Dessins

Suppose that a fixed dessin (S, D) is given. We are now in position to create its associated discrete Belyĭ pair.

Recall that the canonical triangulation $\mathcal{T} = \mathcal{T}(D)$ may not be simplicial. For this and several other practical reasons we perform a barycentric subdivision of $\mathcal{T}$ and denote the resulting complex by $\mathcal{K} = \mathcal{K}(D)$. (See β in Figure 4.1.)

DEFINITION. *Given a dessin* (S, D), *the Circle Packing Theorem implies existence of the Riemann surface* $|\mathcal{K}(D)|_{cp}$, *homeomorphic to* S, *supporting a circle packing* $\mathcal{P}_\mathcal{K}$ *for* $\mathcal{K}$. *We denote this Riemann surface by* s_D *and refer to it as the* **coarse circle packing surface** *for* D *and refer to the packing* $\mathcal{P}_{\mathcal{K}(D)}$ *as the* **coarse circle packing** $\mathcal{P}_D$.

Observe that $\mathcal{T}$ is represented in $\mathcal{K}$: its 0-, 1-, and ∞-vertices are among the vertices of $\mathcal{K}$; each edge of $\mathcal{T}$ is a chain of edges (at this stage, two) of $\mathcal{K}$; and each "dessin" face is now a union of faces (six) of $\mathcal{K}$. Thus s_D constitutes the "surface" part of a discrete Belyĭ pair and provides a geometry for $\mathcal{T}$. When s_D

is the Riemann sphere, a normalization of $\mathcal{P}_D$ is needed. For this purpose, we will designate 0-, 1-, and ∞-vertices v_0, v_1, v_∞ of $\mathcal{T}$ and apply an automorphism to ensure that these points of carr($\mathcal{P}_D$) are located at $0, 1$, and ∞, respectively, on $\mathbb{P}$.

For the meromorphic function part we must first describe the standard coarse circle packing $\mathcal{Q}$ for the Riemann sphere: Choose three points on a topological circle on a topological sphere; this breaks the circle into three segments and defines a triangulation $\mathcal{T}$ of the sphere, the two faces sharing a common boundary. Its barycentric subdivision $\beta\mathcal{T}$ will be denoted by $\mathcal{H} = \mathcal{H}_0$ and defines an abstract simplicial triangulation of the topological sphere. The Circle Packing Theorem guarantees a packing $\mathcal{Q}$ for $\mathcal{H}$ on the Riemann sphere $\mathbb{P}$. This $\mathcal{Q}$ is shown on the right in Figure 3.1, with a normalization placing the vertices of $\mathcal{T}$ at $0, 1, \infty$ and with shading of the triangle of $\mathcal{T}$ (six faces of $\mathcal{H}$) forming the lower half-plane. (The conventional orientation of $\mathbb{P}$ in $\mathbb{R}^3$ means that the "lower" half-plane is the left hemisphere.)

We are now in position to define a discrete meromorphic function $b_D : \mathcal{P}_D \longrightarrow \mathcal{Q}$. The vertices of $\mathcal{T}$ have been decomposed into sets V_0, V_1, and V_∞. Identify each circle associated with a 0-vertex to the circle of $\mathcal{Q}$ centered at 0, each 1-vertex to the circle of $\mathcal{Q}$ centered at 1, and each ∞-vertex to the circle of $\mathcal{Q}$ centered at ∞. Tangencies and orientation clearly dictate that the seven circles of $\mathcal{P}_D$ defining any (+)triangle of $\mathcal{T}$ are carried to the seven circles of $\mathcal{Q}$ forming the unshaded upper half-plane, while the seven forming any (−)triangle are carried to the seven forming the shaded lower half-plane.

DEFINITION. *Given a dessin (S, D), the associated **coarse discrete Belyĭ pair** is the pair (s_D, b_D) consisting of the coarse circle packing surface for D and the discrete meromorphic function $b_D : S_D \longrightarrow \mathbb{P}$.*

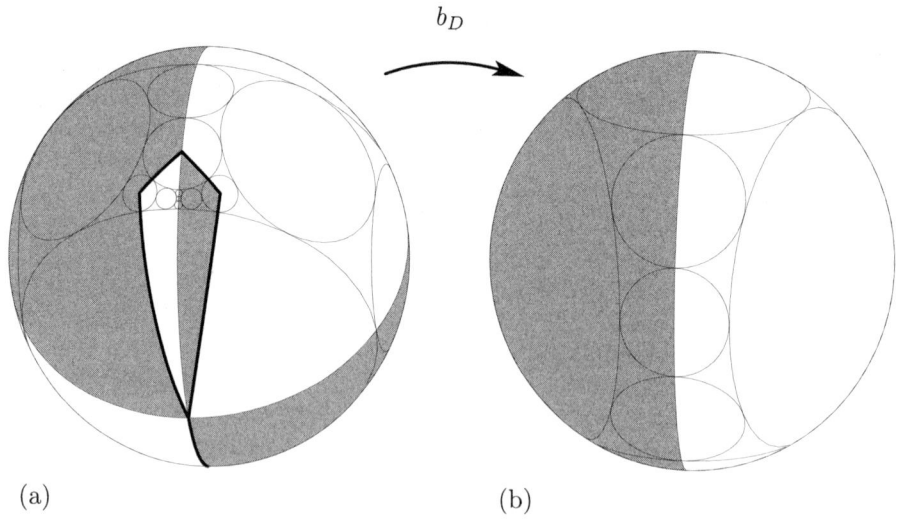

FIGURE 3.1. A coarse discrete Belyĭ pair.

The coarse pair for the simple dessin of Figure 1.1 is illustrated in Figure 3.1. The surface s_D is just $\mathbb{P}$ in this genus 0 case, but the locations of its vertices are determined by the packing $\mathcal{P}_D$, and these define the dessin itself, shown with the thick line.

All nontrivial discrete Belyĭ maps will have branch points. For instance, Figure 3.1 is associated with the dessin of Figure 1.1; it has an ∞-vertex which belongs to 6 faces of $\mathcal{T} = \mathcal{T}(D)$, alternately (+) and (−)triangles. These six are mapped by b_D to the two triangles forming the upper and lower half-planes, respectively, in $\mathcal{Q}$; consequently, this ∞-vertex is a branch point of order two for b_D. This behavior, mimicking the classical situation, is typical: a dessin vertex v which belongs to $2n$ faces of $\mathcal{T}$ engenders a branch point of order $n-1$ in the Belyĭ map. Moreover, one can easily check that b_D does not branch at vertices of $\mathcal{K}$ which are not dessin vertices (i.e., vertices of $\mathcal{T}$). In a clean dessin, all the 1-vertices are simple branch points (order 1).

It is important to recognize that the discrete Belyĭ pair (s_D, b_D) and the classical Belyĭ pair (S_D, B_D) are *qualitatively* indistinguishable. The surfaces s_D and S_D are homeomorphic to one another, since each is homeomorphic to the topological surface S. In fact there exist homeomorphisms $\phi : S \longrightarrow s_D$ and $\Phi : S \longrightarrow S_D$ which respect the embedded $\mathcal{T}$'s. The maps $s_D \circ \phi : S \longrightarrow \mathbb{P}$ and $S_D \circ \Phi : S \longrightarrow \mathbb{P}$ are open, continuous, orientation preserving maps sharing valence and branch structures. In fact, $s_D \circ \phi$ and $S_D \circ \Phi$ are isotopic maps. Topologically and combinatorially, these pairs can't be distinguished.

3.3. Hexagonal Refinement

The discrete objects created so far are termed "coarse" only because they involve so few circles for carrying the geometric information. There is a rule-of-thumb in circle packing: *the finer the circle packing, the closer its geometric behavior to classical conformal behavior*. So we need packings with more numerous and smaller circles if we hope to approach continuous behavior. We deploy a process called hexagonal-refinement.

We use α to indicate the **hex subdivision** operator which, when applied to a complex $\mathcal{K}$ triangulating a surface S, yields the complex $\alpha\mathcal{K}$ also triangulating S. Its effect is illustrated in Figure 3.2: one first adds a vertex to the middle of each edge of $\mathcal{K}$ and then adds three edges in each face of $\mathcal{K}$ to connect pairs of the new vertices. Thus, each triangle (face) of $\mathcal{K}$ is broken into four triangles in $\alpha\mathcal{K}$. The vertices of $\mathcal{K}$ remain, with their degrees unchanged; each new vertex, being in the edge of two faces, has degree six, hence the adjective "hexagonal". Repeated applications will be indicated with powers, $\alpha^n \mathcal{K}$; three stages of hex refinement of a triangle are shown in the figure.

In the case of a dessin (S, D), we have the complex $\mathcal{K} = \beta\mathcal{T}$ triangulating S, the complex $\mathcal{H}$ triangulating the sphere, and the coarse discrete Belyĭ map $b_D : \mathcal{P}_D \longrightarrow \mathcal{Q}$ between their packings. We simply hex refine $\mathcal{K}$ and $\mathcal{H}$, obtain circle packings for the refinements, and arrive at a "refined" map between them.

In particular, by the Circle Packing Theorem, there is a Riemann surface $|\alpha\mathcal{K}|_{cp}$ homeomorphic to S which supports a circle packing for $\alpha\mathcal{K}$; denote the Riemann surface by $s_D^{(1)}$ and the packing by $\mathcal{P}_D^{(1)}$. Likewise, there is a circle packing on $\mathbb{P}$ for the refinement $\alpha\mathcal{H}$, to be denoted $\mathcal{Q}^{(1)}$ and normalized to keep the original three vertices at $0, 1, \infty$, respectively. It is elementary to see that b_D, as a simplicial map between $\mathcal{K}$ and $\mathcal{H}$, induces a simplicial map between $\alpha\mathcal{K}$ and $\alpha\mathcal{H}$; in fact, on the vertex set, the new map extends the original. This defines the circle packing map

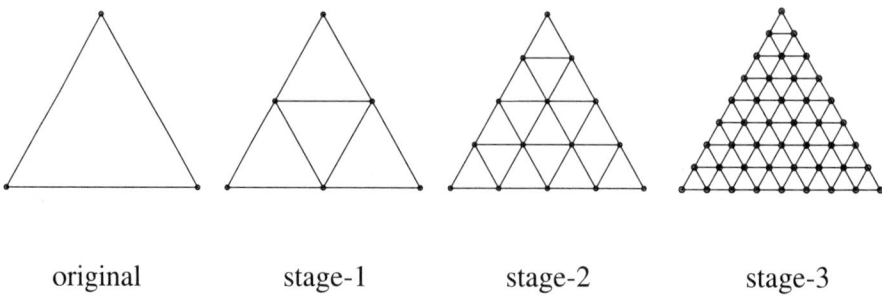

FIGURE 3.2. The hexagonal subdivision operation.

from $\mathcal{P}_D^{(1)}$ to $\mathcal{Q}^{(1)}$ which we denote by $b_D^{(1)}$. We therefore arrive at a new discrete Belyĭ pair $(s_D^{(1)}, b_D^{(1)})$. We may repeat this refinement process any finite number of times.

DEFINITION. *Given a dessin (S, D) and $n \geq 0$, the **stage-n discrete Belyĭ pair** associated with (S, D) is the pair $(s_D^{(n)}, b_D^{(n)})$ obtained from the coarse pair (s_D, b_D) by n successive stages of hexagonal refinement followed by computation of the circle packings. The associated stage-n complex is $\mathcal{K}_n$, the circle packing of $s_D^{(n)}$ is $\mathcal{P}_D^{(n)}$, the stage-n packing of the sphere is $\mathcal{Q}^{(n)}$, so $b_D^{(n)} : \mathcal{P}_D^{(n)} \longrightarrow \mathcal{Q}^{(n)}$.*

It will be a standing assumption in the genus 0 case that $\mathcal{P}_D^{(n)}$ has been normalized as follows: for each $i \in \{0, 1, \infty\}$ an i-vertex v_i has been designated; a Möbius transformation has been applied to the sphere so that vertex v_i of $\text{carr}(\mathcal{P}_D^{(n)})$ is located at i, for $i = 0, 1, \infty$.

A stage-n discrete Belyĭ pair will again be *qualitatively* indistinguishable from the other stages and from the classical Belyĭ pair. In particular, the branch points of $b_D^{(n)}$ occur at dessin vertices (the original vertices inherited from $\mathcal{T}$) and branch orders are preserved under refinement (because degrees of vertices are preserved). The $b_D^{(n)}$ have identical branch structures, and, as maps from S to $\mathbb{P}$, will be isotopic to one another. One would expect the pairs to be distinct, however, in their *quantitative* structure. A packing $\mathcal{P}_D^{(n+1)}$ is not a direct "refinement" of $\mathcal{P}_D^{(n)}$ (no one has discovered a good way to refine circle packings as one could, say, squares), only the *combinatorics* are being refined. In fact, in positive genus cases there is, in the absense of special symmetries, no reason to expect the Riemann surfaces $s_D^{(n+1)}$ and $s_D^{(n)}$ to be conformally equivalent; in other words, the geometries are different.

Figure 1.3 in the Introduction is the stage-3 discrete Belyĭ pair associated with the coarse pair of Figure 1.1. It is intuitively clear that finer packings will carry more geometric information than their coarser predecessors. We make this very precise in Chapter 4.

3.4. Geometric Lemmas

The bounded geometry associated with circle packings is critical to our uniformization work. It is a standard feature of the theory dating from the seminal work of Rodin and Sullivan. In the following, (S, D) is our dessin, as usual; however, the results apply equally well to the more general triangulations of the next section.

The bounded geometry enters primarily through the Ring Lemma [43]. In the following, $d = \deg(\mathcal{K}_n)$, and for $n \geq 1$, $d = \max\{6, \deg(\mathcal{K})\}$.

LEMMA 3.1. *There exists a constant $\Theta = \Theta(d) > 0$, depending only on d, so that the following holds: If S has positive genus, then the angles of the faces of $\operatorname{carr}(\mathcal{P}_D^{(n)})$ are bounded below by Θ for all $n \geq 0$. If S has genus 0, then these angles are bounded below by Θ for n sufficiently large.*

PROOF. In the euclidean case, the Ring Lemma tells us that in a univalent packing, the ratio of the radii of two neighboring (interior) circles is bounded by a constant depending only on degree. This easily converts to a lower bound on angles of faces in the carrier, since they are formed by triples of neighboring circles. The arguments extend easily to cover the hyperbolic setting.

On the sphere, however, the Ring Lemma fails; the ratio of radii for neighbors may be made arbitrarily large by applying a suitable Möbius transformation to the packing. Our normalization becomes crucial: we show in Lemma 3.4 below that for sufficiently large n, the spherical radii of circles of $\mathcal{P}_{\mathcal{K}_n}$ will be bounded by $\pi/6$. In particular, any flower of circles will lie in a hemisphere. Stereographic projection of that hemisphere to the plane, application of the euclidean Ring Lemma, and some elementary geometry shows that the angles of faces of $\operatorname{carr}(\mathcal{P}_D^{(n)})$ will be bounded below by a positive constant depending only on d. □

To convert the bounded geometry of packings to the distortion of associated maps, we need this geometric lemma, whose proof is standard. References to the literature on quasiconformal maps will be given in the next section.

LEMMA 3.2. *Let T and T' be geodesic triangles, each lying in the sphere, the euclidean, or the hyperbolic plane. Assume $\theta > 0$ is a lower bound for the angles of T and for the angles of T', measured in their respective geometries. Then the barycentric map $g : T \longrightarrow T'$ is κ-quasiconformal for a $\kappa \geq 1$ depending only on θ.*

Recall that as point mappings, the discrete Belyĭ map $b_D^{(n)}$ carries faces of $\operatorname{carr}(\mathcal{P}_D^{(n)})$ barycentrically onto faces of $\operatorname{carr}(\mathcal{Q}^{(n)})$. By Lemmas 3.1 and 3.2, the restriction of $b_D^{(n)}$ is κ-quasiconformal on open faces. By removability of analytic arcs and isolated singularities, $b_D^{(n)}$ is κ-quasiregular. In conclusion, we have the following:

PROPOSITION 3.3. *There exists a constant $\kappa = \kappa(d) > 1$ depending only on d so that the following holds: If S has positive genus, then $b_D^{(n)} : s_D^{(n)} \longrightarrow \mathbb{P}$ is κ-quasiregular for all $n \geq 0$. If S has genus zero, then $b_D^{(n)}$ is κ-quasiregular for all sufficiently large n.*

Our last result of the section relies on the Rodin and Sullivan Length-Area Lemma along with the Ring Lemma. It confirms one's intuition that the radii of circles in refined packings should go to zero. We actually require only the genus 0 case in this paper (for use in the proof of Lemma 3.1 above). However, the idea of the proof easily generalizes and may be useful in other circumstances.

When S has genus 0, we require the standard normalization of packings noted earlier. When S has genus 1, the euclidean metric of $s_D^{(n)}$ is determined only up to a scalar, so we may assume that $\text{diam}(s_D^{(n)}) = 1$. No normalization is needed when S has genus greater than one.

LEMMA 3.4. *Let r_n be the maximum radius among the circles of the normalized packing $\mathcal{P}_D^{(n)}$. Then $r_n \longrightarrow 0$ as $n \to \infty$.*

PROOF. We begin on the sphere. Everything follows from very local considerations, so fix a vertex v of $\mathcal{T}$. Let γ denote the simple closed edge-path in $\mathcal{K}$ running through the neighbors of v. In $\mathcal{K}$, this path separates v from at least two of the three designated vertices v_0, v_1, v_∞ used in the normalization of the packings. Without loss of generality, assume γ separates v from v_1 and v_∞.

As $\mathcal{K}$ is *hex* refined, the new complexes $\mathcal{K}_n$ have additional chains of vertices layered between v and γ (which is also refined); in a sense, this region combinatorially "fattens" as n grows. For each n let p_n denote the collection of circles from $\mathcal{P}_D^{(n)}$ corresponding to vertices of $\mathcal{K}_n$ separated from 1 and ∞ by γ and let p_n' be the same collection with the circle $c_{v,n}$ for v removed. Observe that the combinatorics of p_n' depend only on d and n; it is a combinatorial annulus of particularly simple type which we describe below. If $A_n \subset \mathbb{P}$ denotes the annulus bounded by $\partial c_{v,n}$ and γ, then $p_n' \subset A_n$. Below we will use p_n' to establish a lower bound M_n on the modulus of A_n which goes to infinity with n. In other words, $c_{v,n}$ is seen to be separated from 1 and ∞ on $\mathbb{P}$ by an annulus whose modulus goes to ∞ with n. We may conclude by standard arguments about extremal annuli that the spherical radius of $c_{v,n}$ must go to zero, with rate governed by d. This will conclude the genus 0 case.

Since we don't have the Ring Lemma on the sphere, we project p_n' to the plane. Observe that the circles of p_n' break naturally into n disjoint chains of circles. Let the corresponding edge-paths be denoted $\gamma_1, \gamma_2, \cdots, \gamma_n$, ordered so that γ_j separates v from γ_{j+1}. Let $l_1, l_2, \cdots, l_{n+1}$ denote the combinatorial lengths of these paths. γ_1 is just the edge-path through the neighbors of v in $\mathcal{K}_n$, so $l_1 = \deg(v, \mathcal{K}) \leq d$. We obtain easy bounds on the successive combinatorial lengths because these vertices, all resulting from *hex* refinement, are of degree six. In particular, one can show that $l_j \leq j, j = 1, \cdots, n$. That allows us to write down the quantity pertinent to the Length-Area Lemma, ([**43**, p. 353]); namely define $k_n = k_n(d)$ by

$$k_n(d) = \left[\frac{1}{d}(1 + \frac{1}{2} + \frac{1}{3} + \cdots + \frac{1}{n})\right]^{-1/2} \geq \left[\frac{1}{l_1} + \frac{1}{l_2} + \cdots + \frac{1}{l_n}\right]^{-1/2}.$$

Since the harmonic series diverges, k_n goes to zero as n grows.

The argument of the Length-Area Lemma is only indirectly about moduli of ring domains; we need to do some adjustment. First, stereographically project p_n (which avoids ∞) to a packing q_n of the plane. Repack q_n as a "maximal" or Andreev packing $\widehat{q}_n$ in $\mathbb{D}$ with the circle c corresponding to v at the origin (see [**4**]). The chains of circles associated with the γ_j in p_n now correspond to chains of circles separating c from the boundary of the unit disc. By the argument in the proof of the Length-Area Lemma, radius$(c) \leq k_n$. As n grows, c gets smaller and with a little help from the (euclidean) Ring Lemma one can show carr$(\widehat{q}_n')$ nearly fills the annulus $\mathbb{D}\backslash c$. Thus for sufficiently large n we have

$$\log(1/k_n) \leq 2\text{Mod}(\text{carr}(\widehat{q}_n')).$$

3.4. GEOMETRIC LEMMAS

The Ring Lemma and Lemma 3.2 above imply

$$\text{Mod}(\text{carr}(\widehat{q}'_n)) \leq \kappa \text{Mod}(\text{carr}(q'_n)).$$

Under stereographic projection to $\mathbb{P}$, this latter modulus does not change; the image of $\text{carr}(q'_n)$ is a subset of A_n, so finally,

$$\text{Mod}(\text{carr}(q'_n)) \leq \text{Mod}(A_n).$$

Set $M_n = \frac{2}{\kappa}\log(1/k_n)$. This depends only on d, $M_n \to \infty$ as $n \to \infty$, and $M_n \leq \text{Mod}(A_n)$, as desired. With this, we are done with the spherical case.

The positive genus cases are minor variations. In genus 1, the packing p_n may be assumed, with an appropriate lift under the covering map, to lie in $\mathbb{C}$ with $c_{v,n}$ at the origin and lying (by our diameter normalization) within $\mathbb{D}$. The argument of the Length-Area Lemma directly implies that radius$(c_{v,n})$ goes to zero. When genus is greater than one, p_n lifts to $\mathbb{D}$ and we can put $c_{v,n}$ at the origin. Since p'_n separates $c_{v,n}$ from the unit circle, the euclidean radius of $c_{v,n}$ goes to zero as before; since it is centered at the origin, the same applies to its hyperbolic radius. The bounds depend only on d, so we are done. □

Ever finer control of the geometry emerges under refinement, as we will see in the next section when we apply Rodin and Sullivan's Hexagonal Packing Lemma.

Let's end the chapter by observing the geometry of packings at cone points. Figure 3.3 shows four barycentrically subdivided, twice hex refined, circle packed equilateral triangles. In the combinatorics, this is a closed chain of faces, as indicated by the identification arrow, but of course it can't be isometrically embedded in the plane because of the cone angle $4\pi/3$ at the center. Circle packing these combinatorics provides the flat embedding. The resulting map is the discrete analogue of the power map $z \mapsto z^{3/2}$. The dashed lines in the carriers are shown to help in comparing this map to the classical version of Figure 2.1. Note that the "packing" algorithm for computing radii is where the combinatorics and the geometry do battle with one another to achieve this embedding; see the Appendix.

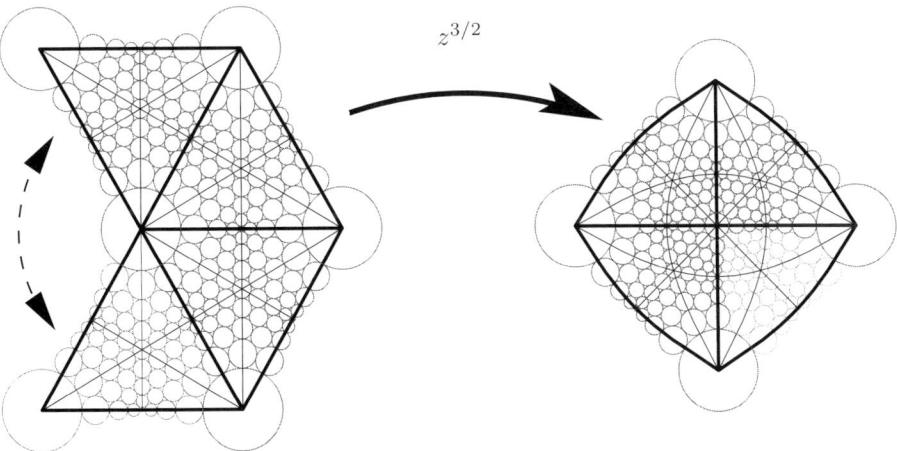

FIGURE 3.3. A discrete power map.

CHAPTER 4

Uniformizing Dessins

In this, the central theoretical section of the paper, we connect the classical and discrete Belyĭ pairs. Our work will apply to general triangulations. In the classical setting we introduce the notion of "reflective surfaces" and their refinement by "conformal subdivision". We then show that it is a small step from a refined classical structure to a refined circle packing structure.

4.1. Reflective Structures and Conformal Subdivisions

The "equilateral" dessin surfaces are examples of what we will refer to as **reflective triangulations**. We work only with triangulations in this paper, but since similiar ideas have been used recently in other settings, we offer a more general definition.

DEFINITION. *A regular cell decomposition $\mathcal{C}$ of a Riemann surface R is **reflective** if, for every edge e of $\mathcal{C}$ which lies in the boundary of two closed faces (2-cells) $f_\pm$, there exists an **anticonformal reflection** of $f_+ \cup f_-$ across e; explicitly, there exists an idempotent homeomorphism r_e that is anticonformal on $\text{int}(f_+ \cup f_-)$, $r_e(f_\pm) = f_\mp$, the restriction of r_e to e is the identity on e, and r_e exchanges the vertices of $f_\pm$ that do not lie on the edge e. In particular, all cells have the same number of vertices. We say that $\mathcal{C}$ endows R with a **reflective structure**.*

In general, the decomposition of a topological surface S into topological n-gons implies existence of a unique conformal structure in which the decomposition is reflective and the cells are conformally regular n-gons, as in §2.2. In [**8**], the plane was endowed with a reflective pentagulation, a refective structure based on pentagonal cells, and work on cubulated manifolds suggests uses for reflective squarings.

REMARK. The conformal reflection r_e across the edge e is not globally defined and, indeed, generically will not extend anticonformally beyond its domain of definition $f_+ \cup f_-$. Even in a covering surface, the reflection does not necessarily extend, as is required, e.g., in Wolfart's "mirror-invariant" triangulations in [**57**]. Though r_e is necessarily angle-preserving at the vertices of $f_\pm$ forming the endpoints of e, it generally will not preserve angles at the remaining vertices of f_+ and f_-. Notice that if v is a vertex of $\mathcal{C}$ of degree d, then the angle between consecutive edges of $\mathcal{C}$ incident at v as determined by the complex structure on R is precisely $\frac{2\pi}{d}$.

EXAMPLE. Let (S, D) be a dessin with classical Belyĭ map $B_D : S_D \longrightarrow \mathbb{P}$. Recall our notation that $S_D = S_\mathcal{T}$, where $\mathcal{T} = \mathcal{T}(D)$ is the canonical triangulation for D. Then $\mathcal{T}$ is a reflective triangulation of S_D: indeed, for any edge e, let B_e be the restriction of B_D to $\text{int}(f_+ \cup f_-)$, where $f_\pm$ are the faces of $\mathcal{T}$ containing e. Then r_e is the continuous extension to $f_+ \cup f_-$ of $B_e^{-1} \circ c \circ B_e$, where c denotes complex conjugation.

PROPOSITION 4.1. *Let $\mathcal{C}$ and $\mathcal{C}'$ be combinatorially isomorphic, reflective cell decompositions of the respective Riemann surfaces R and R'. Then any (orientation-preserving) combinatorial isomorphism of $\mathcal{C}$ with $\mathcal{C}'$ may be realized as a conformal isomorphism of R with R' that takes $\mathcal{C}$ to $\mathcal{C}'$.*

The elementary proof, based on the Schwarz Reflection Principle and removability of isolated singularities, is left to the reader. We state a useful corollary:

COROLLARY 4.2. *Let $\mathcal{T}$ be a reflective triangulation of the Riemann surface R. Then $R_\mathcal{T}$, the Riemann surface determined by an equilateral metric structured on $\mathcal{T}$, is conformally equivalent to R.*

A triangulation $\mathcal{T}$ of an oriented topological surface not only imposes a conformal structure on the surface, but simultaneously realizes itself as a reflective triangulation *in situ* on the resulting Riemann surface. In our method of uniformizing equilateral surfaces we rely crucially on the fact that $\mathcal{T}$ admits arbitrarily fine subdivisions, what we call *refinements*, which themselves form reflective triangulations of this very same Riemann surface.

A **subdivision** of the triangulation $\mathcal{T}$ of the topological surface S is a triangulation $\mathcal{T}'$ of S such that each cell (vertex, edge, or triangle) of $\mathcal{T}'$ is contained in some cell of $\mathcal{T}$. We are interested in subdivisions of a very special type.

DEFINITION. *If $\mathcal{T}'$ subdivides the triangulation $\mathcal{T}$ of the oriented topological surface S, then $\mathcal{T}'$ is a **conformal subdivision** of $\mathcal{T}$ if there is a subdivision of the reflective triangulation $\mathcal{T}$ of the Riemann surface $S_\mathcal{T}$ that is itself reflective in $S_\mathcal{T}$ and combinatorially isomorphic to $\mathcal{T}'$.*

LEMMA 4.3. *If $\mathcal{T}'$ is a conformal subdivision of $\mathcal{T}$, then the associated equilateral surfaces $S_\mathcal{T}$ and $S_{\mathcal{T}'}$ are conformally equivalent. Further, if $\mathcal{T}'$ and $\mathcal{T}$ are realized simultaneously as reflective triangulations of this Riemann surface with $\mathcal{T}'$ subdividing $\mathcal{T}$, and if e is an edge of $\mathcal{T}'$ contained in the edge E of $\mathcal{T}$, then the conformal reflection r_e is the restriction of the conformal reflection r_E to the two faces of $\mathcal{T}'$ contiguous along e.*

The first assertion of the lemma is important in uniformization, since it will allow us to replace the Riemann surface $S_\mathcal{T}$ by $S_{\mathcal{T}'}$, realized as the equilateral surface $|\mathcal{T}'|_{eq}$. However, "conformality" involves an equally important but separate issue: $\mathcal{T}'$ and $\mathcal{T}$ must be realizable simultaneously as reflective triangulations in that common Riemann surface, with $\mathcal{T}'$ subdividing $\mathcal{T}$ *in situ*. As a cautionary example to keep in mind, every triangulation $\mathcal{T}$ of a topological 2-sphere induces the same conformal structure, namely, that of the Riemann sphere $\mathbb{P}$. However, for a generic subdivision $\mathcal{T}'$ of $\mathcal{T}$, the *in situ* triangulations $\mathcal{T}'$ and $\mathcal{T}$ will be incompatible.

PROOF OF 4.3. Since $\mathcal{T}'$ is a conformal subdivision of $\mathcal{T}$, there is a subdivision of the reflective triangulation $\mathcal{T}$ of the Riemann surface $S_\mathcal{T}$ that is itself reflective and combinatorially isomorphic with $\mathcal{T}'$. An application of Proposition 4.1 guarantees that the Riemann surfaces $S_\mathcal{T}$ and $S_{\mathcal{T}'}$ are conformally equivalent, and the first assertion of the lemma follows. The second assertion follows from the fact that conformal reflections are uniquely determined in a neighborhood of their fixed point sets by those very fixed point sets. □

We are interested primarily in barycentric subdivision and the hexagonal refinements described earlier. These easily are seen to be conformal subdivisions, but for completeness we prove the next result, which provides many additional examples.

4.1. REFLECTIVE STRUCTURES AND CONFORMAL SUBDIVISIONS

PROPOSITION 4.4. *Let τ be an abstract triangulation of a triangle t that has combinatorial dihedral symmetry. Let $\mathcal{T}'$ be a subdivision of $\mathcal{T}$ such that the restriction of $\mathcal{T}'$ to each face of $\mathcal{T}$ is combinatorially equivalent to τ. Then $\mathcal{T}'$ is a conformal subdivision of $\mathcal{T}$.*

PROOF. Denote the vertices of t as u_0, u_1, u_2 and recall that $|\tau|_{eq}$ denotes t equipped with an equilateral metric structured on τ. Identifying t with a euclidean unit equilateral triangle $\triangle$, there is a unique homeomorphism $\lambda \colon |\tau|_{eq} \to \triangle$ which fixes u_0, u_1, u_2 and is conformal on the interior of $|\tau|_{eq}$ (recall that the interior of $|\tau|_{eq}$ inherits a conformal structure compatible with its equilateral metric via charts associated with pairs of contiguous equilateral faces, as in §2.2). By replacing the abstract triangulation τ of t with the image triangulation $\lambda(\tau)$, we may assume, without loss of generality, that τ is a reflective triangulation of the equilateral triangle $\triangle$ that is invariant under the action of the dihedral isometry group $D_{2 \cdot 3}$ of $\triangle$. This means, of course, that each edge e of τ that does not lie in the boundary $\partial \triangle$ determines a conformal reflection r_e across e of its two contiguous faces.

In the Riemann surface $S_{\mathcal{T}}$, realized concretely as the equilateral surface $|\mathcal{T}|_{eq}$, let $\mathcal{T}_\tau$ denote the triangulation formed by isometrically mapping the reflective triangulation τ of $\triangle$ onto each face of $\mathcal{T}$ (each a euclidean unit equilateral triangle). The fact that τ is $D_{2 \cdot 3}$-invariant guarantees that this gives a well-defined triangulation. We claim that $\mathcal{T}_\tau$ is a reflective triangulation of $S_{\mathcal{T}}$. Indeed, every edge e of $\mathcal{T}_\tau$ whose interior lies in the interior of a face of $\mathcal{T}$ admits a conformal reflection r_e of its contiguous faces by construction. Further, if e is an edge of $\mathcal{T}_\tau$ contained in an edge of $\mathcal{T}$, then the facts that τ is $D_{2 \cdot 3}$-invariant and $\mathcal{T}$ itself is reflective in $S_{\mathcal{T}}$ imply the existence of a conformal reflection r_e. In fact, if the edge e of $\mathcal{T}_\tau$ is contained in the edge E of $\mathcal{T}$, then, because of the dihedral symmetry across the altitudes of the equilateral triangles, we have $r_e = r_E |(f_+ \cup f_-)$, where $f_\pm$ are the faces of $\mathcal{T}_\tau$ contiguous along e. Thus $\mathcal{T}_\tau$ is a reflective triangulation of $S_{\mathcal{T}}$ that subdivides $\mathcal{T}$ and, hence, $\mathcal{T}'$, which is combinatorially equivalent to $\mathcal{T}_\tau$, is a conformal subdivision of $\mathcal{T}$. □

DEFINITION. *If τ and $\mathcal{T}$ are as in the proposition, $\tau \mathcal{T}$ denotes a subdivision (or refinement) obtained by subdividing each face of $\mathcal{T}$ according to the combinatorics of τ.*

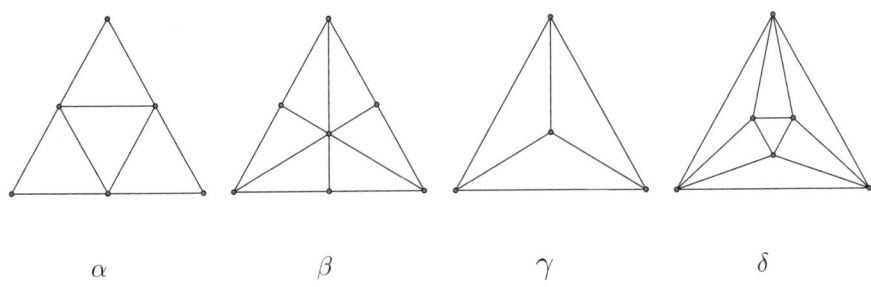

FIGURE 4.1. Conformal subdivision rules.

Figure 4.1 illustrates four $D_{2\cdot 3}$-invariant triangulations of $\triangle$, denoted $\alpha, \beta, \gamma,$ and δ. The first three are conformally correct as shown, since the reflective edges are straight euclidean line segments. The case δ is more generic and its conformally correct form is shown in Figure 4.2; its reflective edges are some unknown curves, neither euclidean line segments nor euclidean circular arcs. (This image of Figure 4.2 was obtained by applying essentially the same methods developed below.)

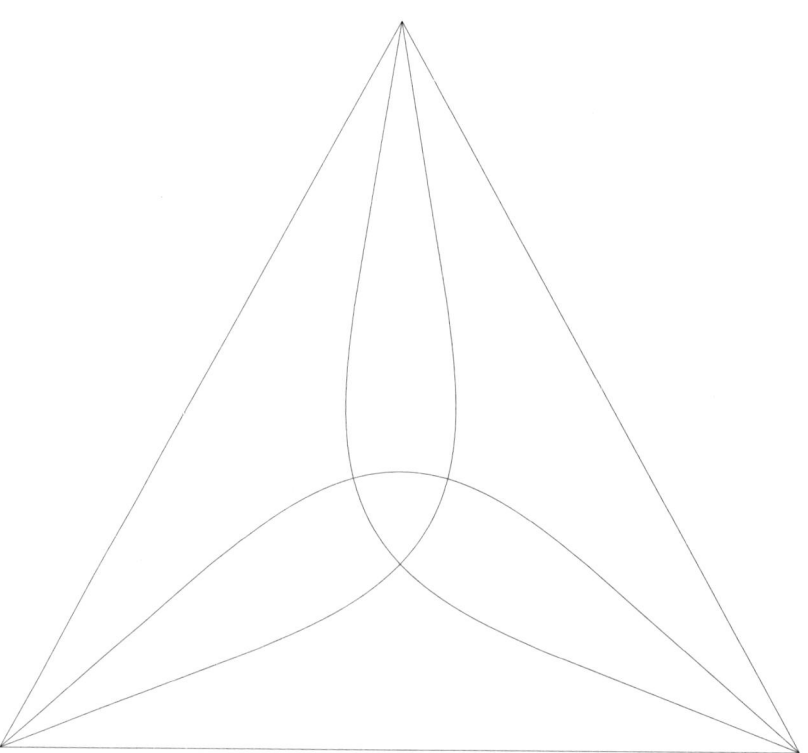

FIGURE 4.2. The reflective subdivision $\lambda(\delta)$.

The uniformization algorithm we develop only requires a sequence of refinements whose meshes go to zero, so it will suffice to work with just the hexagonal and barycentric rules, α and β, respectively. We remind the reader of a trait of *hex* refinement that is of crucial importance; namely, *hex* refinement does not raise degree at the original vertices, whereas the others in Figure 4.1 either double or triple those degrees. We therefore rely on refinement sequences of the form $\{\alpha^n\beta\mathcal{T}\}$, signifying n successive hex refinements of an initial barycentric subdivision of $\mathcal{T}$. The reader will recognize these refinements from Chapter 3. The notation and terminology of refinements apply equally well, of course, to simplicial complexes which triangulate surfaces.

4.2. Uniformizing Equilateral Surfaces

This section establishes the theoretical tools for uniformizing equilateral surfaces, in particular, dessin surfaces. We use many standard results regarding Riemann surfaces and quasiconformal mappings; good references are [33, 21, 34].

Let $\mathcal{T}$ be a triangulation of the compact, connected, oriented topological surface S. Our task is to uniformize the reflective surface $S_\mathcal{T}$ determined by the equilateral surface $|\mathcal{T}|_{eq}$. For practical reasons we work with the barycentric subdivision $\beta\mathcal{T}$. Since barycentric subdivision is conformal, the equilateral surface $|\beta\mathcal{T}|_{eq}$ is in the same conformal class as $S_\mathcal{T}$. By identifying $S_\mathcal{T}$ as the equilateral surface $|\beta\mathcal{T}|_{eq}$ and subdividing each unit equilateral face of $|\beta\mathcal{T}|_{eq}$ into equilateral triangles of side length $\varepsilon(n) = \frac{1}{2^n}$ via n successive hex refinements, we realize $S_\mathcal{T}$ as $|\mathcal{T}_n|_{eq}^{\varepsilon(n)}$, where $\mathcal{T}_0 = \beta\mathcal{T}$ and $\mathcal{T}_{n+1} = \alpha\mathcal{T}_n$ for $n \geq 0$. Thus, $\{\mathcal{T}_n\}$ is a sequence of reflective triangulations of $S_\mathcal{T}$ into triangles, equilateral as seen in the piecewise euclidean metric of $|\mathcal{T}_0|_{eq}$, which are refinements of one another *in situ*, and whose meshes approach zero in the metric of $|\mathcal{T}_0|_{eq}$ as $n \to \infty$.

Note that, unlike the intrinsic metric, the piecewise euclidean metric on the Riemann surface $S_\mathcal{T}$ determined by the triangulation $\mathcal{T}_0$ generally is not conformal. Though it is compatible with the complex structure of $S_\mathcal{T}$ in the sense that it provides analytic charts for $S_\mathcal{T}$ that cover all but the finitely many vertices of $\mathcal{T}_0$, and it is conformal on the complement of these vertices, nonetheless, it fails to be conformal at the non-six degree vertices of $\mathcal{T}_0$. Indeed, though the measure of every angle in every face of $\mathcal{T}_0$ is $\frac{\pi}{3}$ when measured in the euclidean geometry of the equilateral metric, the measure that the Riemann surface $S_\mathcal{T}$ ascribes to a particular angle is $\frac{2\pi}{d}$, where d is the degree in $\mathcal{T}_0$ of the associated vertex.

To aid in the readability of what follows, we continue to use abstract simplicial complexes for the combinatorics on the discrete side of things, as in Chapter 3. Thus $\mathcal{K} = \mathcal{K}_0$ is the simplicial 2-complex which is combinatorially isomorphic to $\mathcal{T}_0 = \beta\mathcal{T}$, and for each $n \geq 1$, $\mathcal{K}_n = \alpha^n \mathcal{K}$ is obtained from $\mathcal{K}$ by n successive hex refinements and is combinatorially isomorphic to $\mathcal{T}_n$.

We are now in position to link the classical and discrete objects. Fix $n \geq 0$. On the classical side we have the Riemann surface $S_\mathcal{T}$ with its reflective triangulation $\mathcal{T}_n$; on the discrete side, the Riemann surface $|\mathcal{K}_n|_{cp}$ determined by the circle packing $\mathcal{P}_{\mathcal{K}_n}$. Since $\mathcal{K}_n$ and $\mathcal{T}_n$ have been canonically identified, we may consider a vertex v of $\mathcal{T}_n$ simultaneously as a vertex of $\mathcal{K}_n$. We now define the key homeomorphisms h_n from $S_\mathcal{T}$ onto $|\mathcal{K}_n|_{cp}$.

DEFINITION. *Let h_n to be a simplicial homeomorphism*
$$h_n : S_\mathcal{T} \longrightarrow |\mathcal{K}_n|_{cp}, n = 0, 1, \cdots$$
from the reflective triangulation $\mathcal{T}_n$ of $S_\mathcal{T}$ to the carrier $\mathrm{carr}(\mathcal{P}_{\mathcal{K}_n})$ *in $|\mathcal{K}_n|_{cp}$ that takes a vertex v in $\mathcal{T}_n$ to the center z_v of the corresponding circle c_v in $|\mathcal{K}_n|_{cp}$ and then extends via barycentric coordinates to map each equilateral triangle uvw in $\mathcal{T}_n$ homeomorphically onto the geodesic triangle $z_u z_v z_w$ in $|\mathcal{K}_n|_{cp}$.*

(When we face the situation in which $|\mathcal{K}_n|_{cp}$ is the sphere, we will have to impose a normalization on $\mathcal{P}_{\mathcal{K}_n}$ to ensure that h_n is unambiguous.)

Our approximation results are nicely formulated in the language of Teichmüller space when S has positive genus. Recall that the Teichmüller space $\mathrm{Teich}(S)$ of the surface S consists of equivalence classes $[g]$ of homeomorphisms g from S onto Riemann surfaces S_g, where the homeomorphisms $g : S \longrightarrow S_g$ and $h : S \longrightarrow S_h$ are equivalent whenever the homeomorphism $h \circ g^{-1}$ is homotopic to a conformal isomorphism of S_g onto S_h. The Teichmüller distance between points $[g]$ and $[h]$ of Teich(S) is

$$\Lambda([g],[h]) = \tfrac{1}{2}\log\inf \kappa(q),$$

where $\kappa(q)$ is the (global quasiconformal) dilatation of q, and q ranges over all quasiconformal mappings in the homotopy class of $h \circ g^{-1}$. This Teichmüller metric Λ is complete, convex, externally convex, and Teich(S) with the Λ-metric topology is homeomorphic to a finite-dimensional euclidean space — to $\mathbb{R}^2$ if S is a genus 1 surface, and to $\mathbb{R}^{6m-6}$ if S is a genus $m \geq 2$ surface. These and other properties of the Teichmüller metric may be found in [**5, 32, 35, 34, 37, 40**].

Now the Riemann surface $S_\mathcal{T}$ is exactly the topological surface S with a maximal complex atlas determined by pairs of contiguous faces from the triangulation $\mathcal{T}$ realized as euclidean equilateral triangles. Thus the identity mapping $\iota_S \colon S \to S_\mathcal{T}$ determines the point $[\iota_S]$ of Teich(S). Further, the homeomorphisms h_n, by forgetting the complex structure on $S_\mathcal{T}$, are mappings defined on S and so provide points $[h_n]$ of Teich(S). We will show that the homeomorphisms h_n are quasiconformal maps with uniformly bounded dilatations; even though those dilatations are bounded away from unity, we have the following result, which shows convergence in moduli of the circle packing surfaces $|\mathcal{K}_n|_{cp}$ to the surface $S_\mathcal{T}$. In many arguments to follow there is a finite set V^* of points of $S_\mathcal{T}$ which must be treated differently; namely, V^* denotes the set of vertices of $\mathcal{T}_0$ that have degree not equal to six.

THEOREM 4.1. *If S has nonzero genus, then in the Teichmüller space* Teich(S),

$$\lim_{n\to\infty} [h_n] = [\iota_S].$$

The pointwise dilatations of the maps h_n are bounded above and converge to unity uniformly on compact subsets of $S_\mathcal{T}\setminus V^*$.

PROOF. We have seen in the results of §3.4 that circle packings induce geometry in a way that controls distortion in a quasiconformal sense. We use this in conjunction with the fact that the Teichmüller metric Λ is proper, meaning that closed Λ-bounded subsets of Teich(S) are compact. This follows, for instance, from the facts that Λ is complete and convex and Teich(S), being homeomorphic to a euclidean space, is locally compact.

By Lemmas 3.1 and 3.2, the restrictions of the homeomorphism h_n to the interiors of the equilateral faces of $S_\mathcal{T}$ are κ-quasiconformal, for some κ depending only on the degree of $\mathcal{T}_0$. By removability of analytic arcs and isolated singularities [**33**], each homeomorphism $h_n : S_\mathcal{T} \longrightarrow |\mathcal{K}_n|_{cp}$ is therefore κ-quasiconformal. It follows that *the sequence $[h_n]$ in* Teich(S) *is contained in the closed Λ-ball of radius $\tfrac{1}{2}\log\kappa$ about the point $[\iota_S]$*.

(The reader might hope that the global quasiconformal dilatations of the h_n converge to unity as $n \to \infty$; however, this generally is not the case. Examine the angle change under h_n at a vertex $a_n \in \mathcal{T}_n$ adjacent a vertex $v \in V^*$. Let f_n be a face of $\mathcal{T}_n$ with v and a_n as vertices. Since $\mathcal{T}_n$ is reflective and f_n is one of exactly six faces that share the vertex a_n, the angle of f_n at a_n has measure precisely $\frac{\pi}{3}$ in $S_\mathcal{T}$, for all $n \geq 1$. On the other hand, the corresponding face f'_n of carr$(\mathcal{P}_{\mathcal{K}_n})$ is asymptotically a euclidean isosceles triangle having angle $\frac{2\pi}{d}$ at v, meaning that its angle at a_n approaches $(d-2)\pi/2d$, where $d = \deg(v, \mathcal{T}_0) = \deg(v, \mathcal{K}_n)$. When $d \neq 6$, this value is not $\frac{\pi}{3}$. Fortunately, such unwanted distortion is restricted to ever-shrinking neighborhoods of the vertices of V^*.)

We shall argue that every convergent subsequence of the sequence $[h_n]$ converges to $[\iota_S]$. This and the fact that Λ is proper then imply that every subsequence of $[h_n]$ has $[\iota_S]$ as a limit point, which in turn implies that the original sequence $[h_n]$ converges to $[\iota_S]$. Let $n(1), n(2), \ldots$ be a strictly increasing sequence of positive integers for which the sequence $[h_{n(i)}]$ converges to the point $[h : S \to S_h]$ of Teich(S). Then there is a sequence of quasiconformal homeomorphisms

$$q_i : |\mathcal{K}_{n(i)}|_{cp} \longrightarrow S_h$$

with q_i homotopic to $h \circ h_{n(i)}^{-1}$ and whose global dilatations $\kappa(q_i)$ converge to unity. We have the following diagram of homeomorphisms, with the upper triangle trivially commuting and the lower triangle commuting up to homotopy.

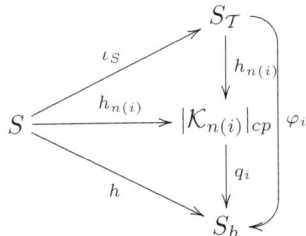

Our aim is to show that the point $[h]$ of Teich(S) is equal to $[\iota_S]$, which will be accomplished by showing that a subsequence of the sequence of quasiconformal homeomorphisms $\varphi_i = q_i \circ h_{n(i)} : S_\mathcal{T} \longrightarrow S_h$ converges uniformly to a conformal isomorphism of $S_\mathcal{T}$ onto S_h that is homotopic to $h \circ \iota_S^{-1}$. Since the maps $h_{n(i)}$ are κ-quasiconformal and the maps q_i have dilatations $\kappa(q_i)$ converging to unity, all but possibly finitely many of the maps φ_i are 2κ-quasiconformal. Standard arguments in [**33**] about covergence of sequences of quasiconformal mappings of plane domains applied in the universal covering surfaces of the surfaces $S_\mathcal{T}$ and S_h imply that there is a subsequence of the sequence φ_i that converges uniformly to a 2κ-quasiconformal homeomorphism $\varphi : S_\mathcal{T} \longrightarrow S_h$. We frame this fact as a separate lemma, which appears, along with its proof, at the end of this proof. Since each map φ_i is homotopic to the fixed map $h \circ \iota_S^{-1}$, it follows that the limit mapping φ is homotopic to $h \circ \iota_S^{-1}$. Modulo the verification of the lemma, our proof that $[h]$ is equal to $[\iota_S]$ is complete once we show that this limit quasiconformal homeomorphism φ is in fact conformal.

Let E be any compact subset of $S_\mathcal{T}$ missing the vertices V^*. Our aim is to show that the dilatation of h_n on E goes uniformly to unity as n grows. Let N be an arbitrary positive integer and choose n so large that each point z of E is centered in a simply connected neighborhood U_z formed by $2N$ generations of the hexagonal grid within the reflective triangulation $\mathcal{T}_n$ and missing V^*.

The set U_z is conformally equivalent to the carrier of $2N$ generations of a *regular* hexagonal circle packing, as illustrated in Figure 4.3(a) for $N = 4$. (In fact, in the equilateral structure $|\beta\mathcal{T}|_{eq}$ of $S_\mathcal{T}$, U_z and this carrier are isometric if the circles are given radius $\frac{1}{2^{n+1}}$.) Write $p_{z,n}$ for the circle packing within $\mathcal{P}_{\mathcal{K}_n}$ corresponding to U_z.

Let us suppose first that S has genus one. Lift $p_{z,n}$ under the covering map of $|\mathcal{K}_n|_{cp}$ to a circle packing $q_{z,n}$ in $\mathbb{C}$. The restriction of h_n to U_z, with image lifted to $\mathbb{C}$, is precisely the circle packing map from $2N$ generations of a *regular* hexagonal circle packing to $q_{z,n}$, as suggested in Figure 4.3. By Rodin and Sullivan's Hexagonal Packing Lemma, [**43**], the maximum pointwise dilatation of h_n in a neighborhood of z is bounded by a quantity depending only on N which goes to unity as $N \to \infty$. Therefore, the maximum of the pointwise dilatations of the restriction of h_n to E converges to unity, as desired.

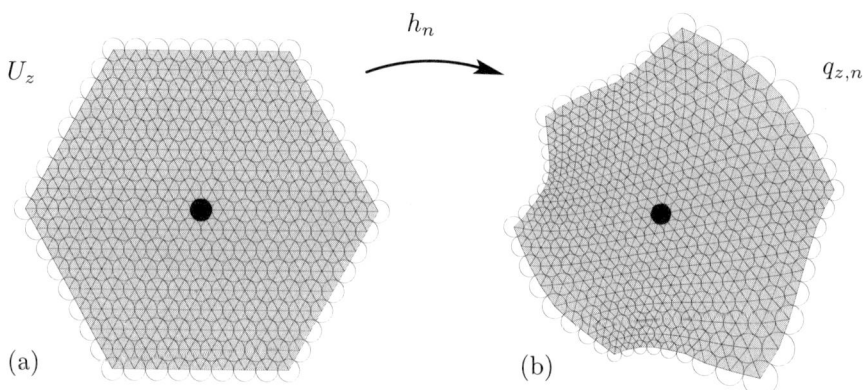

FIGURE 4.3. h_n restricted to $2N$ hexagonal generations.

If S has genus greater than one, our lifted packings $q_{z,n}$ lie in $\mathbb{D}$. However, euclidean geometry is the small scale limit of hyperbolic geometry. By Lemma 3.4, the hyperbolic radii of the circles of $q_{z,n}$ are going uniformly to zero as n grows, so the arguments of the previous paragraph still apply and, again, the pointwise dilatations of the restriction of h_n to E converges to unity.

We have now verified the second part of the theorem. With this it is easy to see that any limit mapping φ of a subsequence of the maps φ_i is conformal. Indeed, since the maximum dilatations $\kappa(q_i)$ converge to unity, the pointwise dilatations of the maps $\varphi_i = q_i \circ h_{n(i)}$ converge to unity uniformly on compact subsets of $S_\mathcal{T} \setminus V^*$. By restricting to simply connected compact subsets of $S_\mathcal{T} \setminus V^*$ and working in the universal covering surface of S_h, we may invoke [**33**, Theorem II.5.3] to conclude that the restriction of the limit mapping φ to any such compact subset is $(1 + \epsilon)$-quasiconformal for every $\epsilon > 0$. It follows that φ is conformal on $S_\mathcal{T} \setminus V^*$ and, by removability of isolated singularities, we conclude that φ is a conformal homeomorphism. This completes the proof of Theorem 4.1, modulo verification of the next lemma. □

LEMMA 4.2. *Suppose R and R' are compact, positive genus Riemann surfaces and φ_i is a sequence of κ-quasiconformal homeomorphisms of R onto R'. Then there is a subsequence of the φ_i's that converges uniformly on R to a κ-quasiconformal homeomorphism of R onto R'.*

PROOF. Let
$$\pi : \mathcal{D} \longrightarrow R \quad \text{and} \quad \pi' : \mathcal{D} \longrightarrow R'$$
be the universal covering projections, where $\mathcal{D}$ denotes appropriately either $\mathbb{C}$ with its usual metric or $\mathbb{D}$ with the Poincaré metric. The groups Γ and Γ' of covering transformations of the respective coverings π and π' are groups of Möbius transformations acting properly discontinuously and by isometries on $\mathcal{D}$. Let $\mathcal{F}'$ be the Dirichlet region for Γ' centered at the origin and, for each i, let z_i be an element of $\mathcal{F}'$ for which $\pi'(z_i) = \varphi_i(\pi(0))$. Let $\Phi_i : \mathcal{D} \longrightarrow \mathcal{D}$ be the lift of the map $\varphi_i \circ \pi$ that sends 0 to z_i, which satisfies $\pi' \circ \Phi_i = \varphi_i \circ \pi$.

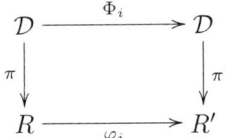

Since the φ_i's are κ-quasiconformal, so too are their lifts Φ_i, to which we may then apply the convergence theorems of [**33**, §II.5]. Our claim is that these convergence results imply the existence of a κ-quasiconformal mapping $\Phi : \mathcal{D} \longrightarrow \mathcal{D}$ that is the limit of a subsequence of the Φ_i's whose convergence is uniform on compact subsets of $\mathcal{D}$. It is a straightforward exercise using the discontinuity of the actions of Γ and Γ' on $\mathcal{D}$ to show that any such limit mapping Φ takes fibers of π to fibers of π' and therefore induces a κ-quasiconformal homeomorphism $\varphi = \pi' \circ \Phi \circ \pi^{-1}$ that is the limit mapping of a subsequence of the φ_i's. Moreover, the convergence of the subsequence of the φ_i's to φ is uniform since the convergence of the subsequence of the Φ_i's to Φ is uniform on compact subsets of $\mathcal{D}$, in particular, uniform on a compact fundamental domain for Γ. The verification of the claimed existence of Φ and its nondegeneracy involves looking separately at two cases, genus one and genus greater than one.

In the genus one, or *parabolic*, case, $\mathcal{D} = \mathbb{C}$. Letting $D = \mathbb{C} - \{0\}$, the restriction of Φ_i to D omits the two values z_i and ∞ of the extended plane $\widehat{\mathbb{C}}$. Their spherical separation is bounded below because $z_i \in \mathcal{F}'$ and $\mathcal{F}'$ is bounded in $\mathbb{C}$, and so the family $\{\Phi_i|D\}$ of κ-quasiconformal mappings of D into $\widehat{\mathbb{C}}$ is normal. It follows that a subsequence of these mappings converges uniformly (in the spherical metric) on compact subsets of D to a limit mapping $\Phi : D \longrightarrow \widehat{\mathbb{C}}$. Since the points $z_i = \Phi_i(0)$ are all in the compact set F', by passing to a further subsequence if necessary, we may assume that the points z_i converge to a point $w \in \mathcal{F}'$, allowing us to define $\Phi(0) = w$. Thus a subsequence of the maps Φ_i converges to $\Phi : \mathbb{C} \to \widehat{\mathbb{C}}$. There are exactly three possibilities: the limit function Φ is either a constant mapping, a mapping of $\mathbb{C}$ onto two points of $\widehat{\mathbb{C}}$, or a κ-quasiconformal mapping of $\mathbb{C}$. We leave it to the reader to eliminate the first two possibilities by using the fact that the maps Φ_i cover the maps φ_i.

For higher genus, the hyperbolic case, $\mathcal{D} = \mathbb{D}$, we get normality of the family $\{\Phi_i\}$ for free, and therefore some subsequence converges uniformly on compact subsets of $\mathbb{D}$ to a mapping Φ. The limit function Φ is either a constant mapping of $\mathbb{D}$ onto a boundary point of $\mathbb{D}$ or a κ-quasiconformal mapping of $\mathbb{D}$ onto itself. Since the points $\Phi_i(0)$ all lie in the compact set $\mathcal{F}'$, the first possibility is precluded and we are done. □

The convergence in Theorem 4.1 can be stated in a more concrete form if we apply the reasoning just used with this lemma. As we did there, let $\mathcal{D}$ denote the universal covering surface of $S_\mathcal{T}$ ($\mathbb{D}$ or $\mathbb{C}$), $\pi : \mathcal{D} \longrightarrow S_\mathcal{T}$ the (analytic) universal covering projection, and $\Gamma \subset \mathrm{Aut}(\mathcal{D})$ the covering group. The triangulation $\mathcal{T}$ in $S_\mathcal{T}$ lifts under π to an infinite triangulation $\mathcal{T}^\infty$ of $\mathcal{D}$ invariant under Γ. Within $\mathcal{T}^\infty$ one can identify a finite union $\mathcal{F}$ of triangles which forms a simply connected fundamental domain of the covering. This concrete setting requires normalization: identify vertices $\nu_0, \nu_1 \in \mathcal{F}$ so that $u_0 = \pi(\nu_0)$ and $u_1 = \pi(\nu_1)$ are neighboring vertices of $\mathcal{T}$; we will always apply a conformal automorphism to $\mathcal{D}$ so that ν_0 is located at the origin and ν_1 is on the positive real axis in $\mathcal{D}$. The conformal data pertaining to $S_\mathcal{T}$ is encoded in $\mathcal{F}$, and in particular, in the locations of the boundary vertices of $\mathcal{F}$. From these (and knowledge of the side-pairings of $\mathcal{F}$) one can generate the covering group, which is (up to conjugation) uniquely associated with the point $[\iota_S]$ of Teich(S).

We can carry out the analogous process for each Riemann surface $|\mathcal{K}_n|_{cp}$. For $n \geq 0$, its universal covering surface is again $\mathcal{D}$; we write $\pi_n : \mathcal{D} \longrightarrow |\mathcal{K}_n|_{cp}$ for the covering projection, Γ_n for the covering group. We may identify a fundamental domain in $\mathcal{D}$, call it $\mathcal{F}_n$, which corresponds combinatorially (under our usual identification of $\mathcal{K}_n$ with $\mathcal{T}$) to $\mathcal{F}$; designated vertices $\nu_0^{(n)}, \nu_1^{(n)} \in \mathcal{F}_n$ lie over u_0, u_1, respectively, and we impose the same normalization as before. Note that $\mathcal{F}_n$ is combinatorially just the refinement $\alpha^n \beta \mathcal{F}$. The ordered list of vertices about $\partial \mathcal{F}$ occur (in order) among the vertices of $\partial \mathcal{F}_n$, each μ of $\partial \mathcal{F}$ occurs as a vertex $\mu_n \in \partial \mathcal{F}_n$ so that $v = \pi(\mu) \in \mathcal{T}$ corresponds to $v = \pi_n(\mu_n) \in \mathcal{K}_n$.

Lifting the maps defined earlier from the surface level up to $\mathcal{D}$, one obtains maps H_n defined by

$$H_n = \pi_n^{-1} \circ h_n \circ \pi, \qquad H_n(0) = 0, \qquad H_n(\nu_1) > 0.$$

Each H_n is a homeomorphism of $\mathcal{D}$ with $H_n(\mathcal{F}) = \mathcal{F}_n$. Standard arguments yield the following corollary to Theorem 4.1.

COROLLARY 4.3. *The homeomorphisms $H_n : \mathcal{D} \longrightarrow \mathcal{D}$ converge uniformly on compact subsets of $\mathcal{D}$ to the identify function. In particular, for each boundary vertex μ of $\mathcal{F}$, the sequence $\{\mu_n\}$ of corresponding boundary vertices of $\mathcal{F}_n$ satisfies*

$$\lim_{n \to \infty} \mu_n = \mu.$$

Likewise, for the covering groups,

$$\lim_{n \to \infty} \Gamma_n = \Gamma,$$

where convergence is in the usual topology of $\mathrm{Aut}(\mathcal{D})$.

In the case that our equilateral surface is associated with a dessin (S, D), this corollary tells us that within the universal covering surface one can essentially see the convergence of the discrete dessin surfaces $s_D^{(n)}$ to the classical dessin surface S_D. The reader will find several examples in the Menagerie.

When S is a genus zero surface, Teich(S) is a singleton and the limit result of Theorem 4.1 is automatic, and uninteresting. Here $S_\mathcal{T} = \mathbb{P}$, and the maps $h_n : \mathbb{P} \longrightarrow \mathbb{P}$ are not defined in a conformally invariant fashion. However, our next result shows that appropriate normalizations of the h_n will again be quasiconformal, with dilatations converging to unity off V^*; in other words, we still have rigidity waiting to be exploited.

For purposes of normalization, fix distinct vertices v_0, v_1, v_∞ of the triangulation $\mathcal{T}$. We assume that each packing $\mathcal{P}_{\mathcal{K}_n}$ has been adjusted by a conformal automorphism of $\mathbb{P}$ to put v_i at i; thus $h_n(v_i) = i$, for $i = 0, 1, \infty$. (When $\mathcal{T} = \mathcal{T}(D)$, we have already designated such vertices.)

THEOREM 4.4. *If S has genus zero, then the (normalized) maps $h_n : \mathbb{P} \longrightarrow \mathbb{P}$ converge uniformly to a conformal automorphism as $n \to \infty$, with the pointwise dilatations of the h_n converging to unity uniformly on compact subsets of $\mathbb{P}\backslash V^*$.*

PROOF OF THEOREM. By Lemmas 3.4 and 3.1, for N sufficiently large, each map $h_n, n \geq N$, is κ-quasiconformal. [**33**, Theorem II.5.1] immediately implies that $\{h_n\}$ forms a normal family of maps, and [**33**, Theorem II.5.3] implies that every limit function of this family is κ-quasiconformal. Lemma 3.4 implies that the local geometry of the packings $\mathcal{P}_{\mathcal{K}_n}$ is infinitesimally euclidean. Therefore, in arguments similar to those in the proof of Theorem 4.1, the Hexagonal Packing Lemma implies the uniform convergence of the dilatations to unity on compact subsets missing V^*, and this in turn implies that every limit function of the family $\{h_n\}$ is conformal. If f and g are two such limit functions, then the composition $g \circ f^{-1}$ is a conformal mapping of the Riemann sphere that fixes the three points $0, 1,$ and ∞, hence, $f = g$. This and the normality of the family imply that the maps h_n converge uniformly to a conformal automorphism. $\square$

4.3. Convergence of the Belyĭ Maps

Given a dessin (S, D) and applying Theorem 4.1 to the triangulation $\mathcal{T}(D)$, we conclude that the associated discrete dessin surfaces $s_D^{(n)}$ converge to the classical dessin surface S_D in Teichmüller space. We now establish the convergence, in an appropriate sense, of the discrete Belyĭ maps $b_D^{(n)}$ to the classical Belyĭ map B_D.

First, one should note the **discrete reflective structure** within $|\mathcal{K}_n|_{cp}$ that parallels the classical reflective structure of $S_\mathcal{T}$. This analogue is visually apparent in packing illustrations, such as Figure 1.3 and Figure 3.1 or later illustrations in the Menagerie. For contiguous shaded and unshaded faces, the reflection across the shared edge should simply interchange the carriers of the packings of the two faces; under refinement we would expect this map to become increasingly more anticonformal. We already have in the homeomorphisms h_n the appropriate machinery. The image $h_n(\mathcal{T})$ of the reflective triangulation $\mathcal{T}$ of $S_\mathcal{T}$ provides a triangulation of $|\mathcal{K}_n|_{cp}$. Given an edge $e \in \mathcal{T}$ and contiguous faces $f_\pm$, recall the edge reflection r_e in $S_\mathcal{T}$; the discrete edge reflection is define by

$$r_e^{(n)} = h_n \circ r_e \circ h_n^{-1}.$$

This is an idempotent quasiconformal homeomorphism interchanging $h_n(f_\pm)$ precisely as suggested visually.

Theorem 4.1 (and Theorem 4.4 in the genus 0 case) implies that $r_e^{(n)}$ is nearly anticonformal away from V^*. To be more precise, recall that our definition of the maps h_n begins with the surface $S_\mathcal{T}$ realized concretely as the equilateral surface $|\beta\mathcal{T}|_{eq}$, so the face f_+ is realized in $|\beta\mathcal{T}|_{eq}$ as the union of six equilateral triangles forming a flat euclidean hexagon. The homeomorphism h_n is then obtained only after n iterations of hex refinement. Let $e^{(n)}$ and $f_\pm^{(n)}$ denote the images $h_n(\alpha^n \beta e)$

and $h_n(\alpha^n\beta f_\pm)$ in the circle packing surface $|\mathcal{K}_n|_{cp}$. Let U be a (small) neighborhood of the vertices of $V^* \cap \alpha^n\beta(f_+ \cup f_-)$. There are at most nine such vertices, up to four forming the original vertices of the two faces $f_\pm$ and five (4-degree vertices) forming the barycenters of the original edges of these two faces. Let U_n denote the image $h_n(U)$ and let κ_n denote the maximum dilatation of $r_e^{(n)}$ taken over the set $(f_+^{(n)} \cup f_-^{(n)}) \setminus U_n$. Theorems 4.1 and 4.4 imply the following result.

PROPOSITION 4.1. *For any edge e of $\mathcal{T}$ and neighborhood U as above, the corresponding sequence κ_n of dilatations converges to unity.*

It is important to observe from this result that when numerically uniformizing reflective surfaces, one obtains not only approximate fundamental regions, as in Corollary 4.3, but also a triangulation of that fundamental region which has the combinatorics of $\mathcal{T}$ and approximates the conformally correct reflective triangulation in the surface $S_\mathcal{T}$. In the zero genus case, the numerical uniformization gives a triangulation of the Riemann sphere into nearly reflective triangles whose vertices and edges are close to their conformally correct positions. In the case that $\mathcal{T}$ is the canonical triangulation $\mathcal{T}(D)$ of a dessin D, the numerical uniformization provides the approximate locations of points that branch over $0, 1$, and ∞ under a Belyĭ map, as well as the approximate pre-image of the extended real line under that map. The authors know of no other general method for obtaining such approximations.

Recall from Proposition 3.3 that the discrete Belyĭ maps are κ-quasiregular. The next result guarantees that the Belyĭ map B_D may be approximated as closely as desired by these discrete ones.

THEOREM 4.2. *Given a dessin (S, D) and the associated discrete Belyĭ maps $b_D^{(n)}$, define the maps*
$$\beta_n = b_D^{(n)} \circ h_n : S_\mathcal{T} \longrightarrow \mathbb{P}, \quad n \geq 0.$$
Then the sequence $\{\beta_n\}$ converges uniformly on $S_\mathcal{T}$ to the Belyĭ map B_D. The pointwise dilatations of the maps β_n converge to unity uniformly on compact subsets of $S_\mathcal{T} \setminus V^*$.

PROOF. The proof is similar to those of Theorems 4.1 and 4.4. Fix a $(+)$triangle f of the dessin triangulation $\mathcal{T} = \mathcal{T}(D)$. As in the proof of Threorem 4.4, the Ring Lemma and [**33**, Theorem II.5.1] imply that the restrictions of the maps β_n to f form a normal family of κ-quasiconformal maps, for some constant $\kappa \geq 1$. By [**33**, Theorem II.5.5], every limit function of this family must be either a κ-quasiconformal homeomorphism or a mapping onto a single point in the extended real line. We verify in the next paragraph that the former possibility holds for every limit function. Assuming this for the moment, an argument as in the proof of Theorem 4.4, using the fact that any limit function must take the i-vertex of f to the point i, for $i = 0, 1, \infty$, implies that the restrictions of the β_n's to f converge to a conformal map taking the boundary ∂f to the extended real line $\widehat{\mathbb{R}}$. Immediately, this limit function must be the unique conformal mapping of f onto the upper half plane that takes the i-vertex of f to i. This is precisely the restriction of the Belyĭ map B_D to f. The same argument works if f is any $(-)$triangle. It follows that the maps β_n converge uniformly to the Belyĭ map B_D. The dilatation convergence (already used when the argument of Theorem 4.4 was applied) follows as in the proof of Theorem 4.1.

4.3. CONVERGENCE OF THE BELYĬ MAPS

We now verify that any limit function of the normal family of maps described in the previous paragraph takes at least one value in the (open) upper half plane. Recall the abstract simplicial complex $\mathcal{H}$ triangulating the sphere, defined in Chapter 3; let v denote the vertex corresponding to the barycenter of the upper half plane, one of the two faces of $\mathcal{H}$. Recalling that v is then contained in all the hex refinements $\alpha^n \mathcal{H}$ of $\mathcal{H}$, let $z(n)$ denote the center of the circle c_v in the circle packing $\mathcal{Q}^{(n)}$ for $\mathcal{H}_n = \alpha^n \mathcal{H}$. The definitions of h_n and $b_D^{(n)}$ guarantee that the restriction of each map β_n to the (+)triangle f satisfies $\beta_n(\widehat{z}) = z(n)$, where $\widehat{z}$ is the (unique) conformal barycenter of f. Since the restriction of each map β_n to f is a κ-quasiconformal homeomorphism onto the upper half plane, it suffices to show that the points $z(n)$ are bounded away from the extended real line. In fact, however, the $z(n)$ form a constant sequence: each $z(n)$ is precisely the point $e^{\pi i/3}$, which is the (unique) conformal barycenter of the upper half plane thought of as a triangle with vertices 0,1, and ∞ and sides forming the extended real line. This follows from the fact that the circle packing $\mathcal{Q}^{(n)}$ inherits order six dihedral symmetry from that of $\mathcal{H}_n$; this symmetry permutes the circles centered at 0,1, and ∞ and hence fixes the circle corresponding to the barycenter $\widehat{z}$. $\square$

CHAPTER 5

A Menagerie of Dessins d'Enfants

For convenience, the Menagerie is organized by genus. As opportunities arise, we point out various pertinent features, many of which will apply in several settings. We will refer back to these examples when we discuss computational issues in Chapter 6, but at this point we should remark on the two senses in which these examples are "approximations". First, all are subject to the usual round-off and truncation errors of numerical computation and display; second, even were we to have perfect information on a given discrete dessin, that dessin might only approximate its classical companion.

5.1. Genus 0

In the case of the sphere, the conformal structure engendered by a dessin is not at issue — the sphere has only one. We have latitude only in the normalization, and we have agreed to the convention that designated i-vertices are placed at i for $i = 0, 1, \infty$. In such a normalized situation, the conformally correct dessin, the location of zeros, ones, and poles, and various other metric information are examples of data of interest.

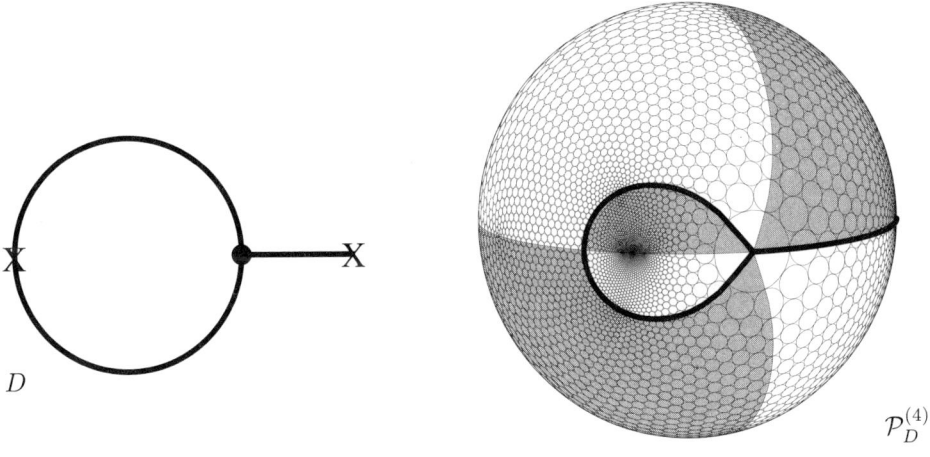

FIGURE 5.1. Dessin 1 and its stage-4 packing in $\mathbb{P}$.

Example 1. We begin with the very simple dessin D of Figure 5.1. This is the only example we give which fails to be "clean"; it has a "free" edge ending in a ×, so its Belyĭ map will not branch at this 1-vertex. Its stage-4 packing $\mathcal{P}_D^{(4)}$ is shown. As will be standard, we shade appropriate faces and mark the dessin itself with a heavy line.

Example 2. We have used a clean dessin only slightly more complicated than the previous example for purposes of illustration earlier in the paper; we denote it Example 2. The dessin and its triangulation are displayed in Figure 1.1, its stage-3 discrete Belyĭ map in Figure 1.3, and its coarse discrete Belyĭ map in Figure 3.1. In Figure 5.2 we project its stage 3 packing to the plane for comparison to Figure 1.1.

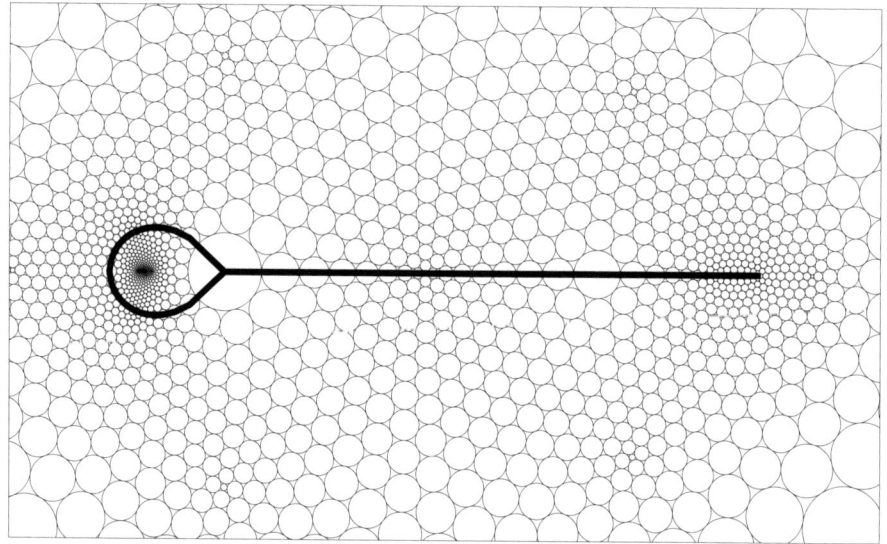

FIGURE 5.2. Stereographic projection of Dessin 2.

Example 3. We add further edges to the dessins of Examples 1 and 2 to get Dessin 3 of Figure 5.3(a). (See §6.1 concerning "dessin moves".) Stereographic projections of the discrete dessin as embedded by the coarse and first four stages of hex refinement are given in Figure 5.4, illustrating the evolving shape. Detail around the head at stage-4 is shown in Figure 5.5.

This dessin, due to Gunter Malle (see [**29**]) has a known and nontrivial orbit. Recall that dessins are associated with number fields. The Galois group of the number field provides an action on dessins, and the Galois orbit of a dessin is the smallest collection of dessins closed under the Galois action. In the case of Example 3, the Galois orbit consists of Dessin 3 and the dessin of Figure 5.3(b) — in other words, the Galois action switches between the left-arm person and her right-arm companion!

Example 4. Trees form the only general class of dessins for which approximation methods have been developed. Dessin 4 is a tree from [**44**]. The embedded

(a) (b)

FIGURE 5.3. Dessin 3: A genus 0 Galois orbit.

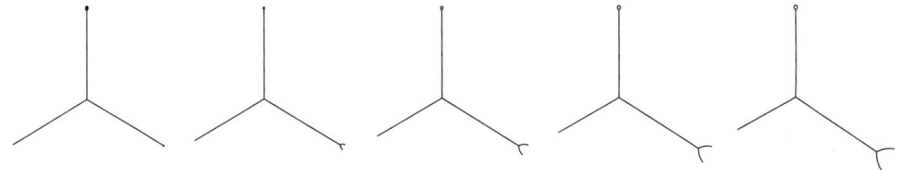

FIGURE 5.4. Successively finer stages of the right-armed dessin.

image of Figure 5.6 is from the stereographic image of $\mathcal{P}_D^{(4)}$ and should be compared to the image [44, p. 112]. Using Grobner basis methods, Couveignes and Granboulan can provide extreme accuracy for embedded trees, enough to eventually recover coefficients of defining equations. (The equation for Dessin 4 has integer coefficients of over twenty digits!) This far surpasses the accuracy available via

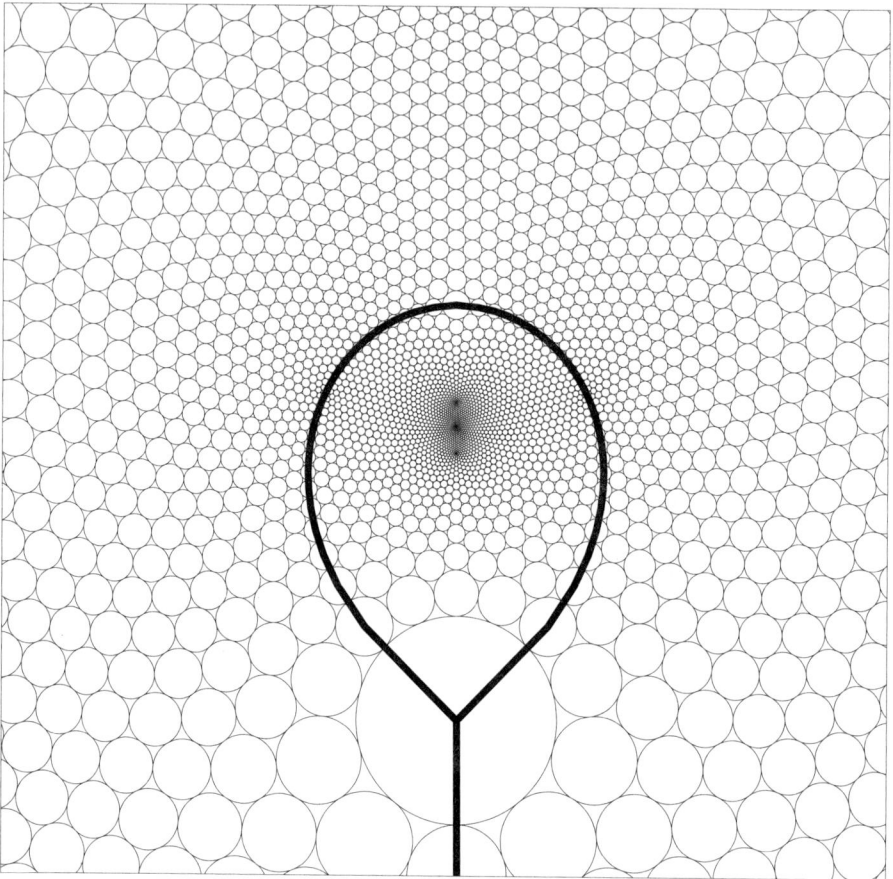

FIGURE 5.5. Stereographic projection of the head of Dessin 3.

circle packing. However, their methods involve certain initial guesses for locations of vertices and the addition of vertices to existing trees. Circle packing might prove very helpful in this process; see especially §6.1.

5.2. Genus 1

A dessin has genus 1 if it is drawn on the topological 1-torus $\mathbb{T}$. The torus has the euclidean plane as its universal cover, and $\mathbb{T}$ endowed with a conformal structure is typically identified with a fundamental domain in $\mathbb{C}$. The lift to the plane of any point on the conformal torus forms a doubly periodic lattice, and the ratios of pairs of complex numbers generating that lattice parameterize the conformally distinct marked tori. We will see this played out in its discrete form here.

We concentrate on a dessin introduced by Shabat and Voevodsky [45] and thoroughly analyzed by them and others. In this case the associated number field, the defining equation, the Galois orbit, and the parameters of the associated tori are known, affording one an opportunity to compare the discrete and classical information.

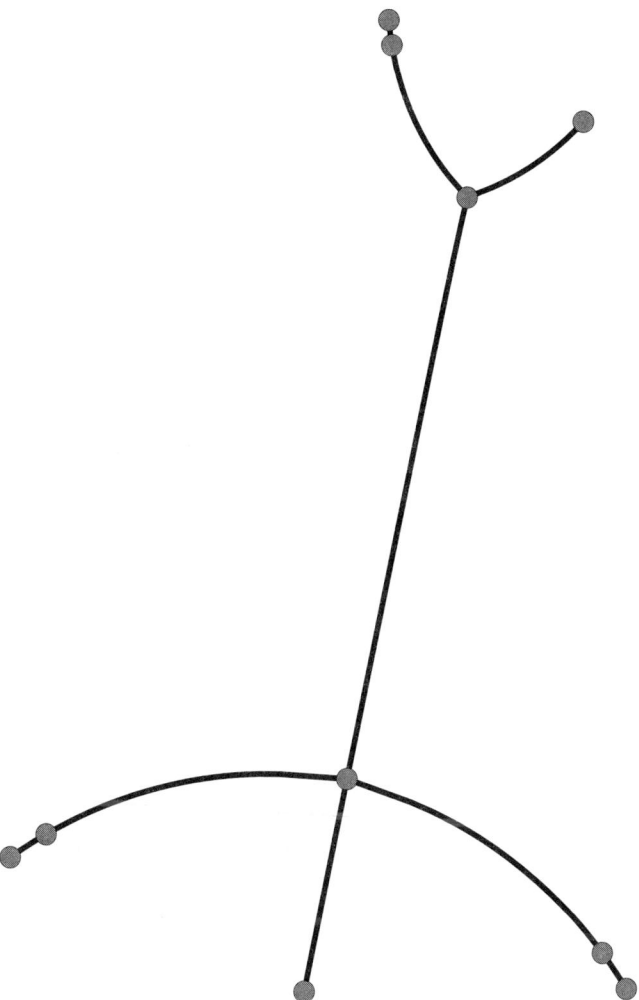

FIGURE 5.6. A dessin tree.

Example 5. Dessin 5 is illustrated in Figure 5.7(a), where opposite sides of the rectangle are identified in the usual way so that the dessin is seen to lie on $\mathbb{T}$.

From Dessin 5 one obtains the canonical triangulation of $S = \mathbb{T}$ and its barycentric subdivision $\mathcal{K}$. There is a unique conformal torus s_D, the coarse dessin surface, which supports the circle packing $\mathcal{P}_{\mathcal{K}}$ for $\mathcal{K}$; in this genus 1 case, the metric is euclidean, hence only defined up to a multiplicative positive constant.

As in the classical setting, one may represent s_D by lifting to the universal covering surface, $\mathbb{C}$. In fact, everything lifts: the triangulation, the complex, the

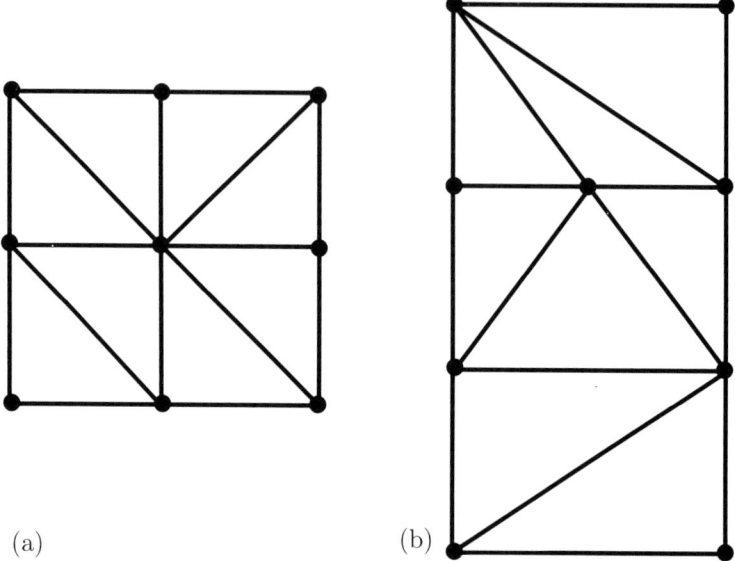

FIGURE 5.7. Dessins 5 and 6: A genus 1 Galois orbit.

circle packing, the metric, and the dessin. The results for the coarse packing of Dessin 5 are illustrated in Figure 5.8(a), with a fundamental domain highlighted.

We are in position to estimate the conformal modulus of the discrete torus. The four corners of the fundamental domain in Figure 5.8(a) correspond to the same point of s_D and define the covering lattice in $\mathbb{C}$. From the packing centers one can read off approximations to a pair of complex numbers which generate the lattice. For convenience we have placed the lower edge of the packing with ends at 0 and 1 in Figure 5.8(a); the complex number associated with the left side is approximately $\tau \approx 0.249612 + 0.968346i$; we will refer to this information in Chapter 6.

Of course we have the option of refining our circle packing for a more accurate approximation of the classical dessin. Using 2 stages of hex refinement and repacking leads to the fundamental domain of Figure 5.8(b) for $s_D^{(2)}$. Again, the lower corners have been placed at 0 and 1, and the complex number for the other side can be read off as approximately $\tau \approx 0.248308 + 0.968683i$.

Example 6. The Galois orbit for Dessin 5 contains one other dessin, shown in Figure 5.7(b). Dessin 6 is laid out using the stage-2 packing $s_D^{(2)}$ in Figure 5.9. This illustrates the general fact (see [**28**]) that conjugate dessins share the same numbers of dessin faces, edges, and vertices. (Their circle packings consequently share the same number of circles at each refinement stage.) Dessins 5 and 6 should be compared, both in their schematic and embedded forms, to [**2**, Fig. 8, p. 209] and [**45**, p. 215].

5.3. Genus 2

With higher genus we move into the hyperbolic realm. The dessin surfaces are covered now by the hyperbolic plane, which we will represent as the unit disc $\mathbb{D}$ endowed with the Poincarè metric $ds = 2/(1 - |z|^2)|dz|$ of constant curvature -1.

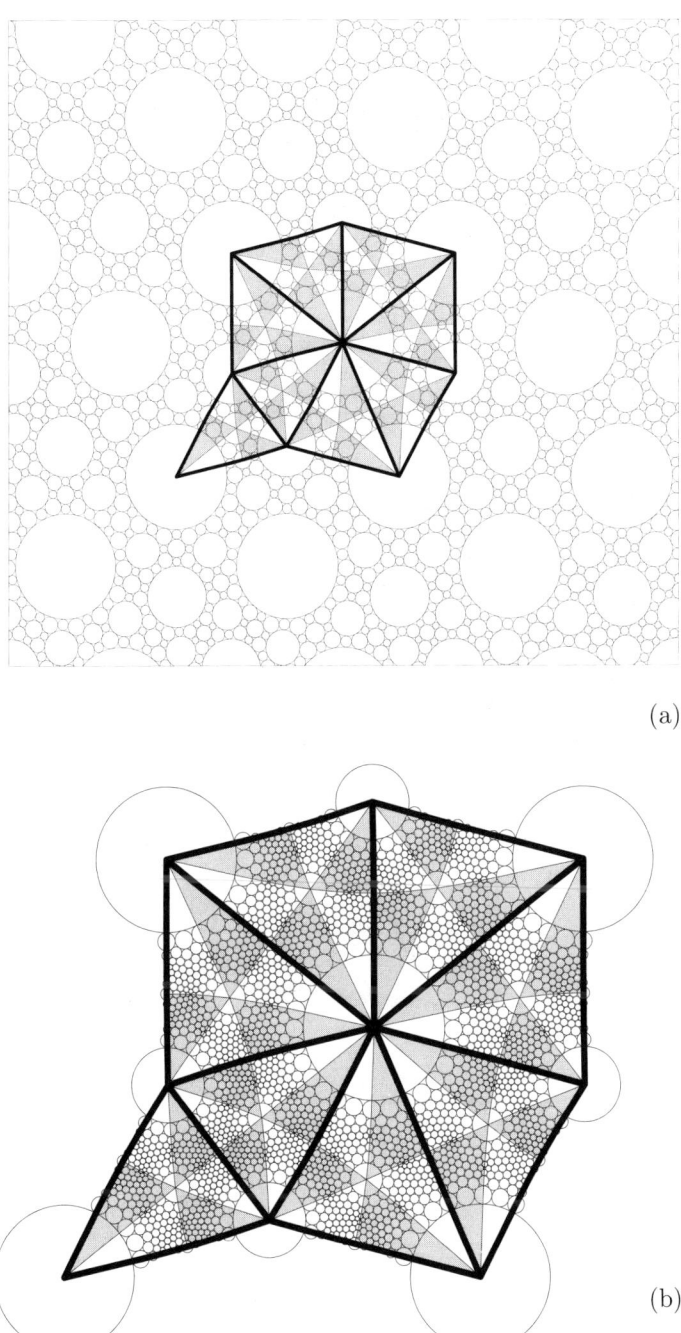

(a)

(b)

FIGURE 5.8. Fundamental domains for genus 1 packings.

FIGURE 5.9. The Galois conjugate of Dessin 5.

In a manner completely analogous to the previous genus 1 setting, one may lift all structures from the surface to a fundamental domain $\mathcal{F}$ of the covering. The covering group Γ is now a discrete nonabelian group of automorphisms of $\mathbb{D}$. We will illustrate with three examples.

Example 7. Dessin 7 is shown in Figure 5.10(a); it is important to note the identifications of the sides which makes this into a surface of genus 2. When triangulated, Dessin 7 contains 28 faces. The associated coarse circle packing is shown

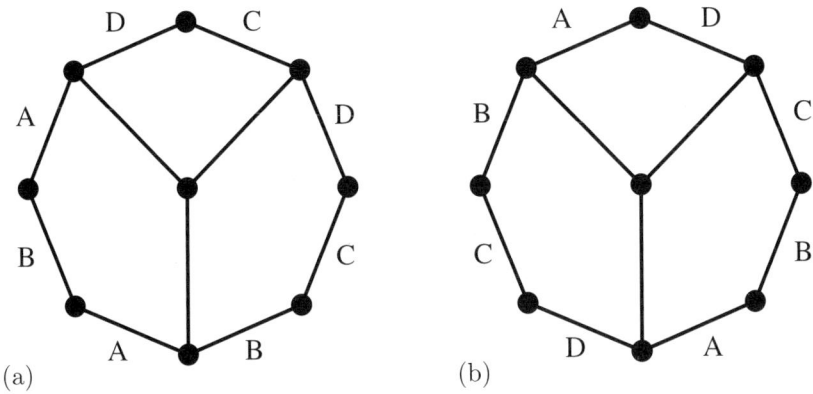

FIGURE 5.10. Dessins 7 and 8: Two dessins of genus 2.

in Figure 5.11(a) and the stage-3 refinement in Figure 5.11(b). In each case, a 0-point has been placed at the origin and only fundamental domains $\mathcal{F}_0$ and $\mathcal{F}_3$, respectively, are shown; the numbering of boundary edges for the side-pairings is indicated on the coarse packing. In each case, the action of the covering group Γ_n generates a tiling of the hyperbolic plane with isometric images of $\mathcal{F}_n$.

A few observations are in order. The eight circles at the cusps on the boundary of $\mathcal{F}_0$ are in fact eight lifts of one circle; in particular, although their radii are euclideanly different because of their distances from the boundary, they share a common hyperbolic radius. The angles at the cusps sum to 2π, and in the fundamental domain for the classical surface would all be precisely $\pi/4$. A cautionary note: the edges between cusps in these figures are approximating analytic arcs; though they may appear to be approximate geodesics, this need not be the case.

This dessin provides an opportunity to illustrate Corollary 4.3. In particular, *the circle packing for a discrete dessin provides numerical estimates of the covering maps of the associated classical dessin surface.* Estimation occurs in two stages which we now discuss. (The same general considerations applied in the euclidean setting of Example 5.)

Step 1: First is the estimation of the covering maps for a discrete dessin surface. In Figure 5.11(a), let $\{c_1, \cdots, c_8\}$ denote the eight cusp points of $\mathcal{F}_0$. Each side-pairing of $\mathcal{F}_0$ is associated with an element $\gamma \in \Gamma_0$ which carries one ordered pair of these points to another: $\gamma : (c_i, c_j) \mapsto (\gamma(c_i), \gamma(c_j))$. This information alone determines the automorphism γ. In turn, such side-pairing automorphisms γ generate Γ_0. Therefore, the locations and pairings of the eight points c_j determine Γ_0. In theory, this information is exact, but in practice, of course, the data is subject to roundoff errors in the computation of packing radii, the location of the centers, and the subsequent computation of the automorphisms.

Step 2: The second stage involves the use of successively finer circle packings so that their covering maps converge to the classical covering maps. Figure 5.11(b) is the stage-3 hex refinement, involving 5374 circles. The computed locations of the eight cusp points give estimates for the associated side-pairing maps. Matrices in $GL(2, \mathbb{C})$ representing the A, B, C, and D side-pairings (see Figure 5.11) are given

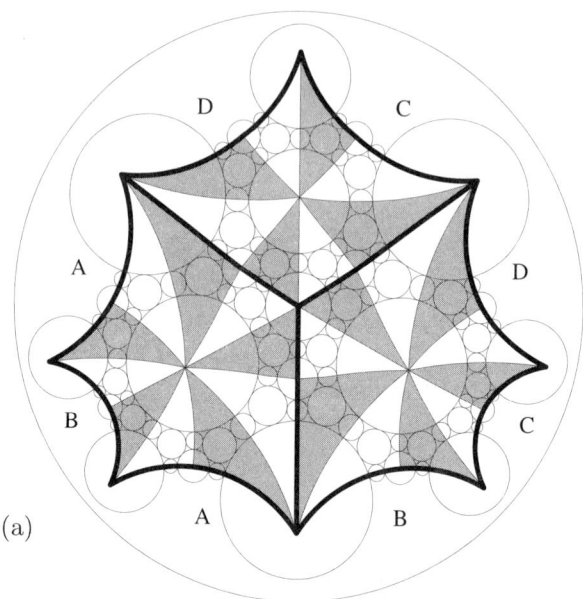

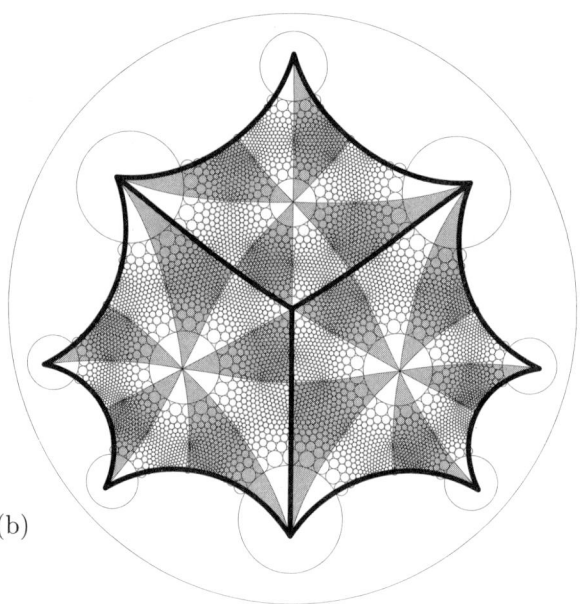

FIGURE 5.11. Coarse and stage-3 fundamental domains for Dessin 7.

by, respectively,

$$\begin{bmatrix} 0.610254 - 0.705705i & -0.725003 + 0.416873i \\ -0.725003 - 0.416873i & 0.610254 + 0.705705i \end{bmatrix}$$

$$\begin{bmatrix} -0.389727 + 0.416402i & 0.511033 + 0.158976i \\ 0.511033 - 0.158976i & -0.389727 - 0.416402i \end{bmatrix}$$

$$\begin{bmatrix} -0.426400 + 0.422296i & -.0443194 + 0.336691i \\ -.0443194 - 0.336691i & -0.426400 - 0.422296i \end{bmatrix}$$

$$\begin{bmatrix} 0.532299 - 0.413640i & 0.605959 - 0.0348912i \\ 0.605959 + 0.0348912i & 0.532299 + 0.413640i \end{bmatrix}$$

It seems clear that estimates of the type obtained here do not provide perfect information — a precise number field, for example — about the dessin surface. However, they do provide what would normally be considered as fundamental information for one to "know" its conformal structure. The data should be sufficient, for instance, to invoke some of the available programs for working numerically with Riemann surfaces, such as the CARS program of the HCM network on Computational Conformal Geometry and the Symbolic Computation Group at Florida State University.

Example 7 is a favorite of the authors because it was the first hyperbolic example attempted: we were surprised with the speed, beauty, and accuracy of the process. The accuracy still seems remarkable, and we comment on this in the next section.

Example 8. Dessin 8, shown in Figure 5.10(b), differs from the previous example in its side-pairings. The coarse packing is shown in Figure 5.12(a), and a visual comparison with Figure 5.11 suggests that these surfaces are conformally very close. Are they the same surface? Is this, in fact, the same dessin?

Example 9. Here we show a somewhat more generic genus 2 dessin; the dessin and its faces, as embedded by a stage-2 packing, are laid out in Figure 5.12(b). One side-pairing has been highlighted to demonstrate that this is clearly not a standard fundamental domain bounded by geodesics.

This dessin was constructed to provide one handle which is simple and one with richer combinatorics. The resulting asymmetry in their conformal structures is evident. (One "handle" consists of the two cells at the top, the other of the nine lower cells.) This highlights a central issue: *To what extent can one anticipate conformal implications directly from the combinatorics?*

5.4. Higher Genera

We begin with two very classical surfaces: the Klein "Hauptfigur" of [22] and the Picard curve. Historically, these resulted from other considerations, and only in hindsight are associated with "dessins". We also display a generic genus 4 dessin to show the level of complexity that our methods can handle. The main hurdle in preparing such examples lies in specifying the dessin combinatorics.

Example 10. Klein's "Hauptfigur" is a genus 3 surface which has played a seminal role in classical geometry and function theory. It is the **modular curve** $X(7)$ of degree seven and has an automorphism group of order $168 = 84(g-1)$, the maximal possible in genus $g = 3$. This curve also has an honored place in the visual

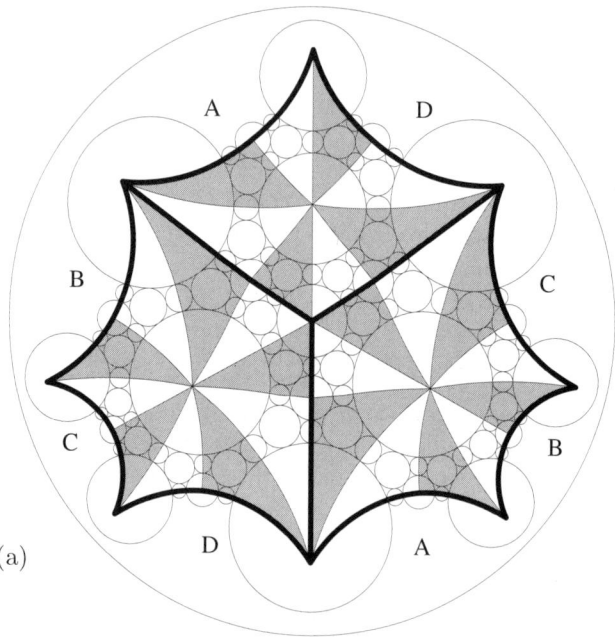

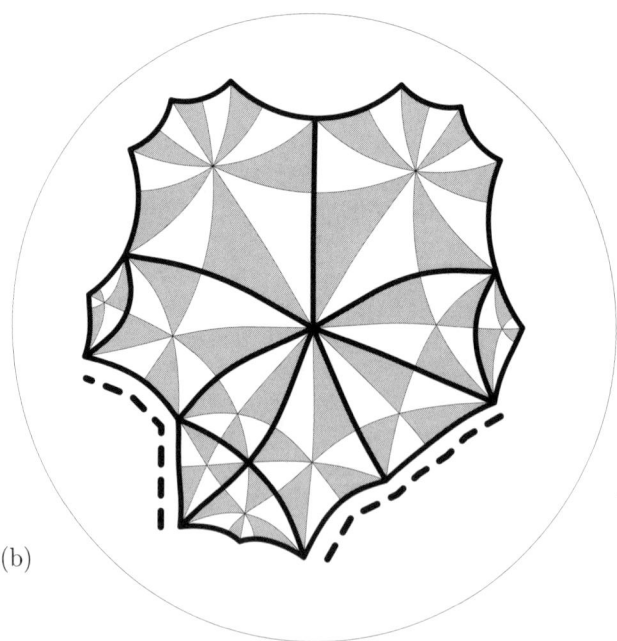

FIGURE 5.12. Dessins 8 and 9.

history, thanks to the beautiful illustrations, now over a century old, produced for Klein.

Klein's surface R is the compact surface $\mathbb{H}/\Gamma(7)$, where $\mathbb{H}$ is the upper half-plane and $\Gamma(7)$ is the subgroup of $PSL(2,\mathbb{Z})$ consisting of elements congruent to $\pm id$ modulo 7. See, for example, [**30, 36**]. A natural triangulation G on R (which one can obtain geometrically, combinatorially, or *via* the group structure) is a 7-regular graph. A classical fundamental domain for R can be indentified within a tiling of the hyperbolic plane by equilateral hyperbolic triangles having angles $2\pi/7$.

We take (R,G) as our dessin; the circle packing (not shown) provides the image in Figure 5.13(a), an image familiar to analysts for more than a century. Figure 5.13(b) shows the associated dessin overlaying the carrier of the same circle packing. For the reader's benefit, this latter illustration indicates the side-pairings associated with the surface.

Note that the dessin faces of the dessin surface are *geodesic* $(2,3,7)$ triangles due to the ubiquitous symmetries of G. In particular, even with the coarse circle packing, these illustrations have perfect accuracy (up to the usual computational roundoff). It is because the dessin faces are geodesic triangles that this picture could be constructed a century ago — this will not be the case in our next examples.

Example 11. Next is the genus 3 "Picard" curve $y^3 = x^4 - 1$. We first employ a dessin taken from Shabat and Voevodsky [**45**, p. 217]; this is pictured with the fundamental domain from the coarse packing in Figure 5.14(a).

The Picard curve can also be constructed by methods similar to Klein's surface, using the fact that it has a very rich automorphism group, one of order 48. In [**41**], J. R. Quine builds the fundamental domain of Figure 5.14(b) using $(2,3,12)$ triangles. The side-pairings are determined by labeling the sides counterclockwise, from 1 (as indicated) to 24; each even-numbered side k is then paired with side $k+7$. It turns out that this triangulation arises from a dessin distinct from the one of Shabat and Voevodsky; the triangles forming the fundamental domain are rearranged in Figure 5.14(c) to more easily picture this alternate dessin. For the side-pairings, number the sides as before from 1 and identify each odd-numbered side k with side $k+5$.

It is convenient to have two genus 3 dessins determining the same conformal structure, since it affords us opportunities to judge the quality of circle packing approximations. (Is there some reason to anticipate this circumstance based solely on the combinatorics of the two dessins?) Comparing the triangulations, note that the triangles of Figure 5.14(b) (and (c)) are, as with the fundamental domain for Klein's surface in Figure 5.13(a), geodesic triangles, so this fundamental domain is essentially exact. On the other hand, the triangles of Figure 5.14(a) may appear to be geodesic, but in fact are not. (One can see, for example, that in reflections of dessin faces across certain edges, the opposite angle on one side will be $\pi/12$, while that on the other is $\pi/4$.) The image of Figure 5.14(a), obtained from a coarse circle packing, is, as far as we know, only approximate.

Example 12. Our final example is a garden variety genus 4 dessin. The main difficulty with higher genera is largely that of encoding the combinatorics. Here we use the "pair of pants" paradigm.

A pair of pants is basically a sphere with three round discs removed; any compact hyperbolic Riemann surface can be decomposed as a disjoint union of a finite

50 5. A MENAGERIE OF DESSINS D'ENFANTS

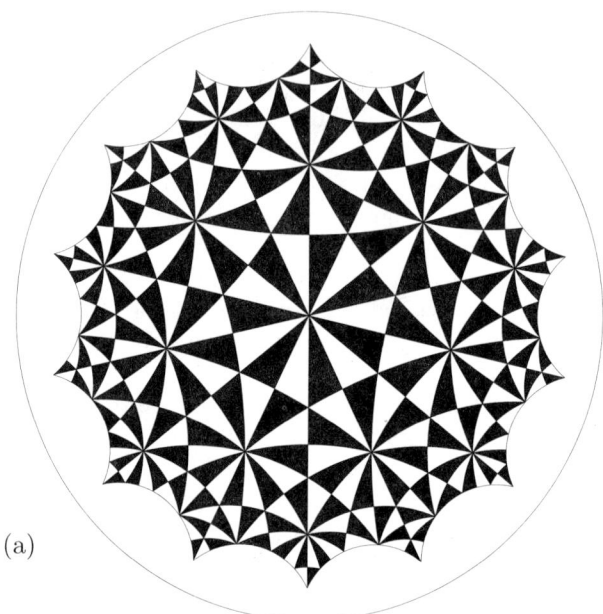

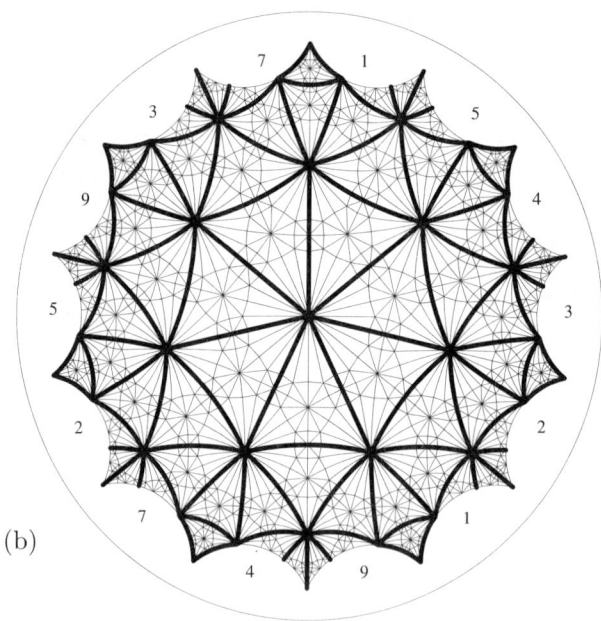

Figure 5.13. Klein's Hauptfigur.

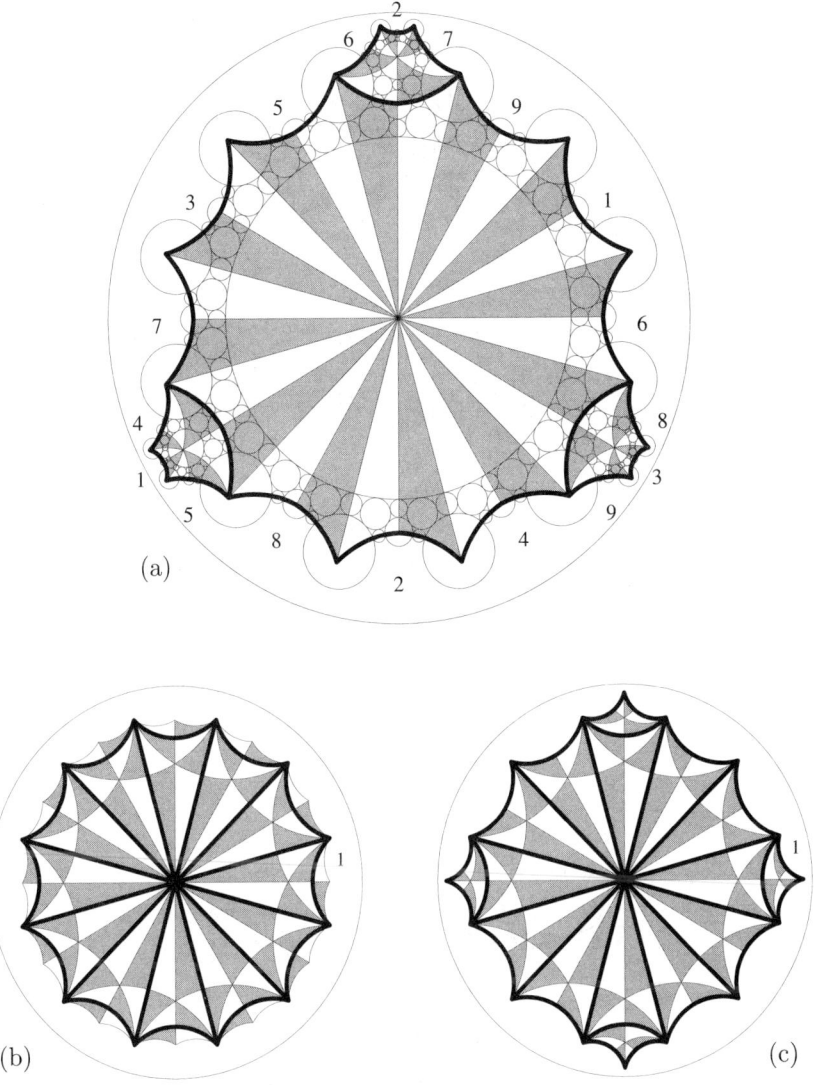

FIGURE 5.14. Two dessins for the Picard surface.

number of pairs of pants (with appropriate structure parameters) identified along boundary components. We will use the basic pair of pants shown in Figure 5.15, with conformal structure determined by two dessin-like 2-cells. A genus 4 surface constructed from 6 identical pairs of pants, twelve 2-cells, is shown in Figure 5.16.

Observe that at this point we have a highly symmetric surface, especially so since the particular pair of pants we are using has the greatest possible symmetry (only countably many surfaces can be constructed from from copies of this pair). We have also built in some global symmetry for easy visualization. (One consequence is, as with Klein's surface, that the discrete conformal structure is precisely equal to the classical.)

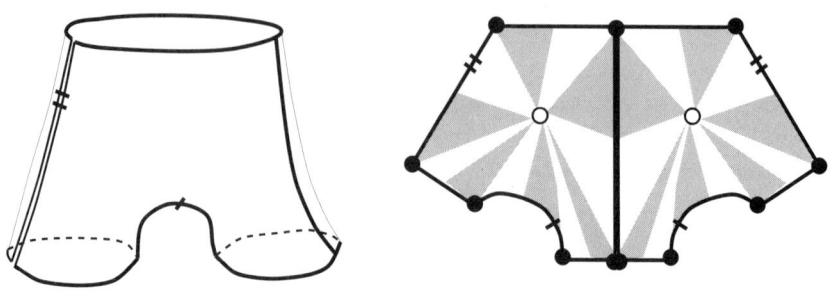

FIGURE 5.15. A standard pair of pants.

FIGURE 5.16. A symmetric genus 4 surface.

6. COMPUTATIONAL ISSUES

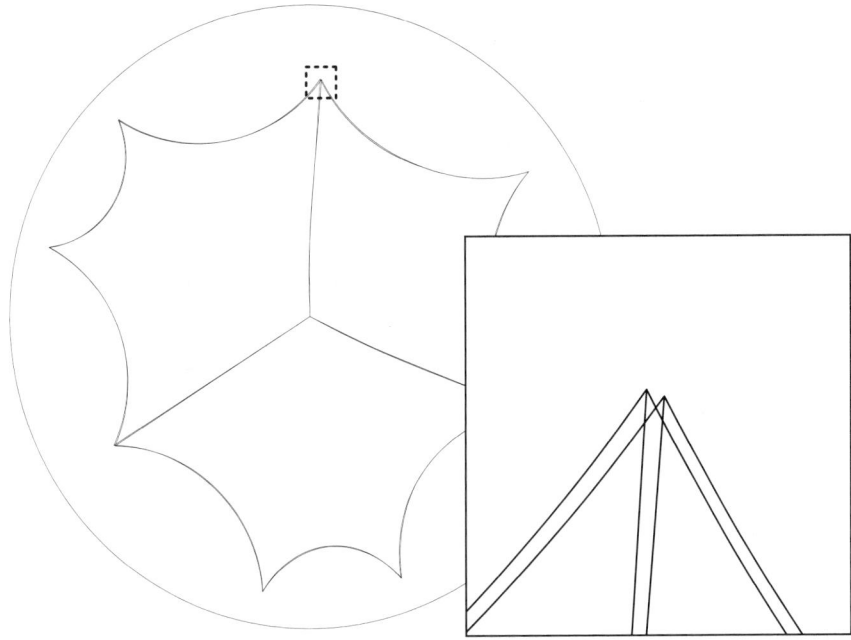

FIGURE 6.1. Comparing cusps for stages of Example 7.

meet with equal angles. In various of our illustrations of discrete dessins one can clearly see that this fails (e.g. Figure 1.3 or Figure 5.5). This is due, in part to our drawing method: dessins are drawn as piecewise geodesic arcs running through the centers of appropriate chains of circles. Accuracy would presumably improve if they were drawn as splines, with the requirement that they run through appropriate sequences of tangency points and come together at vertices with equal angles. This makes little difference in the positive genus cases. However, in spherical geometry, circle centers are not conformally determined (they are not preserved under Möbius transformations); tangency points are, and the use of splines can make noticable differences in dessin drawings. In any case, the pictures in the paper (and the package `CirclePack`) use the more computationally convenient circle centers.

Issue III concerns the accuracy of the descrete vis-a-vis the classical structures. The stability of the discrete conformal structures under refinement in conjunction with Theorem 4.1 suggests some level of accuracy. Of course, further work needs to be done on the numerical side prior to any claims. Nonetheless, the Menagerie provides some opportunities to address accuracy beyond the very symmetric cases such as Klein.

In particular, the genus 1 dessin of Example 5 has been studied thoroughly and its classical j-function value was computed in [**45**, p. 215], namely:

$$4(8+3\sqrt{7})^2(2-\sqrt{7})^6(10-3\sqrt{7})^3 = 457208 - 172564\sqrt{7} \approx 646.5707574.$$

On page 42 we gave numerical estimates of the complex modulus τ at the coarse and at the stage-2 levels for the torus of Example 5. We put these values through *Mathematica* to obtain the estimated j-function values, $j(\tau)$, with these results:

Coarse stage: $635.06238 - 0.00035i$ \hspace{2em} Stage-2: $642.57533 - 0.01011i$,

representing errors of about 1.78 and 0.62 percent at the coarse and stage-2 levels, respectively. (**Remark:** We construct the j-function itself in §7.2.)

There are other opportunities to compare discrete and classical data, as with the tree of Example 4. We could also point to the interesting Picard surface, Example 11. In this instance we have two dessins which define the same surface. For one, the dessin faces are geodesic triangles, so as with the Klein surface, the discrete structures are all identical to the classical one. There seems no reason to expect this for the other dessin, so again one can compare the approximate structures to the known classical one.

The reader may have observed in our examples that circles associated with dessin vertices are often quite large, even in finer stages; one might be led to question the precision. This behavior, however, is wholly consistent with classical behavior. Recall that the number n of faces incident to a dessin vertex v is $2(b+1)$, where b is the order of the branch point B_D has at the point $z \in \mathbb{P}$ associated with v. If $b \geq 1$, B_D compresses large neighborhoods about z into small neighborhoods about $B_D(z)$. In other words, B_D is approximately constant on a large neighborhood of z. This, rather than any loss of precision, is what the large circle at v reflects for the discrete mapping.

The question of theoretical accuracy — how closely the discrete conformal structures (resp. discrete Belyĭ maps) for a dessin approximate the classical conformal structures (resp. classical Belyĭ maps) seems to be very deep. As we point out in the final chapter, approximation is not the only legitimate aim for studying the discrete structures.

6.1. Dessin Modifications

There is a small number of elementary **dessin moves**, reminiscent of Reidemeister moves in knot diagrams, which allow one to construct an arbitrary dessin on a compact surface from any beginning dessin. This is not difficult to establish, and the reader can furnish the details. We mention this because these dessin moves are now not only *practical* in the experimental setting, but almost *essential*. Nearly all of our more complicated dessins have been built from simple seeds by dessin moves which added complexity. In turn, the fact that moves can be implemented opens a new avenue for thinking about dessin structures and their relations to one another.

The basic dessin moves are of three types: (I) adding/deleting vertices, (II) adding/deleting free edges, and (III) adding/deleting separating edges. Recall that dessins are connected and 2-colorable graphs, and the moves must preserve this. So, for example, a type I move adding a vertex actually requires adding two vertices. With this caviat, the moves are largely self-explanatory. It is also useful, particularly working in the category of clean dessins, to identify common "composite" moves, such as adding a double free edge or a loop at a 0-vertex, or adding a "bridge" across a 2-cell from one 0-vertex to another. Only the add versions of these composite moves have been implemented in `DesPack` so far, and these were used in creating the Menagerie.

Concrete examples of moves are illustrated in Figure 6.2. We started with the simplest genus 1 dessin, two loops at a base point, and applied a sequence of "composite" moves leading to Dessin 6, the lower right hand figure. The coarse dessin, circle packing, and shaded dessin faces are shown at various intermediate stages, all normalized to place the bottom corner circles at 0 and 1.

When D' is obtained from the dessin D by a finite sequence of dessin moves, each of which is an "adding" move as opposed to a "deleting" move, we call D' a **dessin refinement** of D.

PROPOSITION 6.1. *Any two dessins D and D' on an oriented closed surface S share a common refinement.*

The proof is left to the reader. It helps to note that by applying a small isotopy to D', one may assume that D and D' are in general position. Their union may then be converted into a 2-colorable graph of which they are both refinements. A corollary to this proposition is: *On a topological surface S, any dessin D' may be obtained from any other dessin D by a finite sequence of elementary dessin moves.*

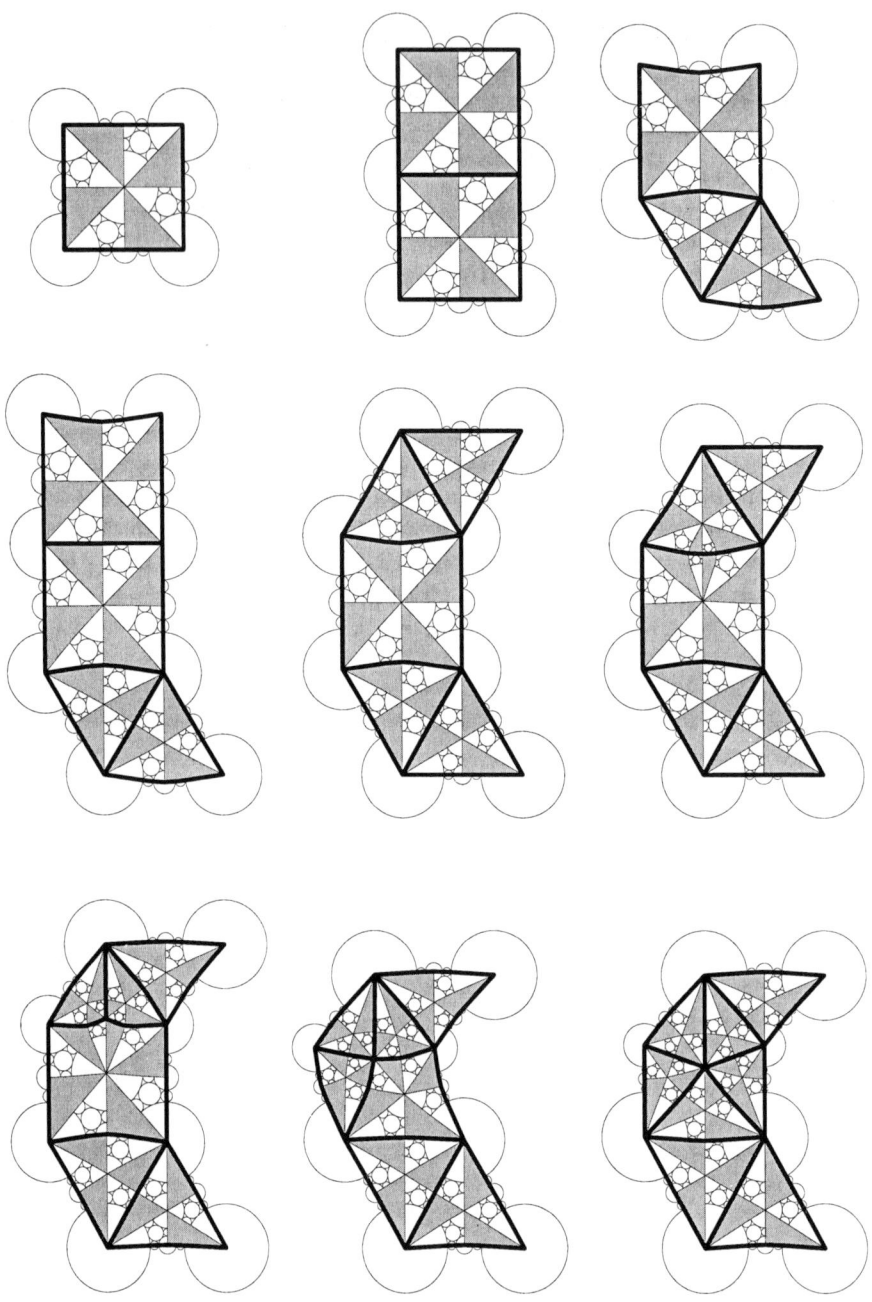

FIGURE 6.2. Dessin moves generating Dessin 6.

CHAPTER 7

Additional Constructions

The piecewise equilateral surfaces at the heart of the theory of dessins are instances of more general *piecewise flat* surfaces, and the associated Belyĭ maps form but a limited special class in the much broader theory of meromorphic functions. In this chapter we will illustrate other settings where *combinatorial* and *conformal* considerations juxtapose in a similar way and where discrete geometry based on circle packing can perhaps play a pivotal role. In each case we find that in special circumstances one can mimic the techniques we've develped earlier, but that interesting and perhaps very difficult open questions remain if one hopes for any generality.

7.1. Conformal Tilings

The conformal construction techniques we have used with triangulations extend to more general polygonal decompositions. This is not just idle generalization; connections between conformal structures and the combinatorics of tilings/shinglings has been under intense study by Jim Cannon, Bill Floyd, and Walter Parry in the context of Thurston's Geometrization Conjecture (see [**10, 11**]). As with dessins, the authors brought circle packing and its experimental capabilities to this topic, leading to the notion of *conformal tilings* in [**8**].

We will illustrate the topic here by recapping the story of the regular pentagonal tiling from [**8**], displayed in Figure 7.1. By blending ideas from dessin constructions with a new rational iteration discovered by Cannon, Floyd, Kenyon, and Parry [**13**] we will see the pleasing conclusion to this tale.

This story began with the *subdivision rule* of Figure 7.2, describing combinatorially how to subdivide a pentagon into a union of 6 pentagons. It's convenient to reverse the arrow in this picture and work with an *aggregation rule*: a rule for pasting 6 pentagons together to form a new pentagon. Aggregation can be repeated *ad infinitum*, leading in the limit to an infinite combinatorial pattern T of pentagons. A geometric realization $|T|$ of this pattern was built by treating each abstract pentagon as a regular euclidean pentagon of unit side length. As with triangles, this defined a locally reflective structure, a piecewise flat metric, and finally a conformal structure on $|T|$. Mapping this Riemann surface conformally to the hyperbolic or euclidean plane, as appropriate, would lead to a conformal tiling $\mathcal{T}$ combinatorially equivalent to T.

At this point in the story, circle packing entered as a means of visualizing $\mathcal{T}$. At the first, coarse, stage a vertex was added to each pentagon of T to obtain an infinite triangulation $\mathcal{K}$. Finite pieces of $\mathcal{K}$ can then be given geometry via circle packing. Figure 7.3 gives two examples, with the heavy edges corresponding to the pentagons. The standard way to improve on the conformal shapes of the pentagons would be

7. ADDITIONAL CONSTRUCTIONS

FIGURE 7.1. A 'regular' pentagonal tiling of the plane.

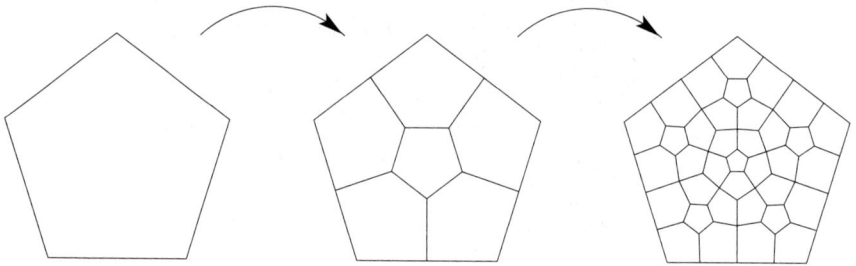

FIGURE 7.2. Pentagonal subdivision rule of Cannon, Floyd, and Parry.

7.1. CONFORMAL TILINGS

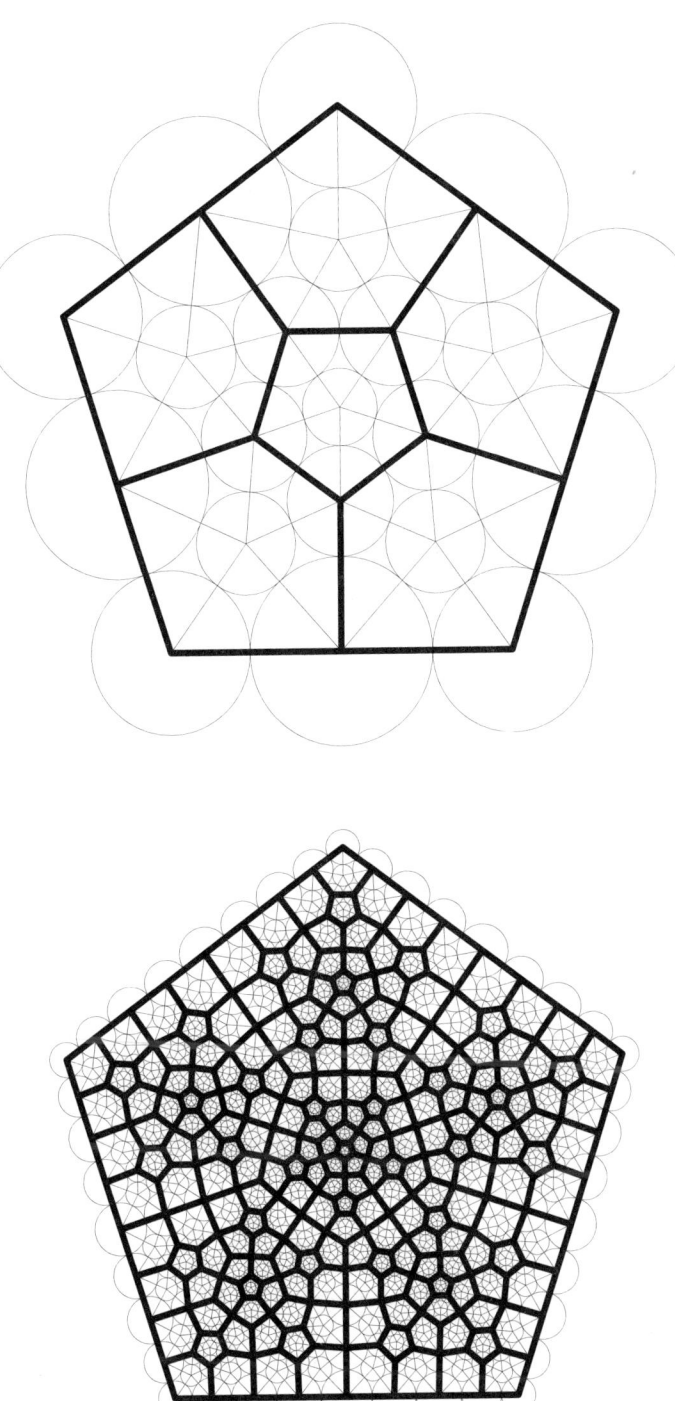

FIGURE 7.3. Initial circle packing embeddings.

to pack successive hex refinements of them. However, the pictures suggested an alternative route: use successively larger pieces of $\mathcal{K}$, but look at aggregates rather than individual tiles. Compare the central pentagon of the pattern on the top in Figure 7.3 to the aggregate pattern of the 36 pentagons around the origin of the pattern on the bottom. Inspired by the evolving shape of such aggregates, we proved by standard conformal geometry arguments the existence of a conformal self-similarity $\phi : \mathcal{T} \longrightarrow \mathcal{T}$ which implements the subdivision rule. This implied simultaneously that $\mathcal{T}$ tiled the euclidean (rather than the hyperbolic) plane and that $\phi \mapsto z \longrightarrow \lambda z$ for some complex scalar λ.

The story concluded in [8] with a numerical estimate of λ based on circle packing images and an open question, inspired by our dessin work: "Is λ an algebraic number?"

Here Cannon, Floyd, Kenyon, and Parry picked up the thread, and at the 1998 Barrett Lectures at the University of Tennessee made a gift to the organizer (the second author) of the exact scalar:

$$\lambda = (-324)^{1/5}.$$

We feel justified in calling this the "pentagonal" number because of its connection with the pentagonal tiling $\mathcal{T}$ coupled with it representation using the first five decimal digits.

So let us see the conclusion of the story and the source of the pentagonal number. Our thanks to Cannon, Floyd, Parry, and Kenyon for allowing us to describe this instance of their recent work [13]. Here's an overview: We begin with our usual decomposition of the sphere into two triangles, the unshaded upper half plane and the shaded lower half plane, with vertex set $\{0, 1, \infty\}$. Then we describe a carefully chosen lifting of this to a more complicated alternately shaded triangulation of the sphere. Putting a conformal structure on this new triangulation in our usual manner leads to the associated Belyĭ meromorphic function; since both domain and range are the Riemann sphere, this is necessarily a rational function $R : \mathbb{P} \longrightarrow \mathbb{P}$. It turns out to be a "post-critically finite" rational map of the type studied by Thurston (see [18]). In particular, with appropriate normalizations, one can exploit combinatorial symmetries to show that lifting our initial triangulation under iterates of R implements stages of the pentagonal subdivision rule for appropriate aggregates of the lifted triangles. The pentagonal tiling $\mathcal{T}$ itself results from a fifth root of a renormalization of this infinitely regressive lifting process. The final step in finding λ involves finding an explicit formula for R.

To describe the combinatorics we need a schematic representation of the sphere. We will find it convenient to slit the sphere along the positive real axis and flatten it to a "kite" shape. Figure 7.4(a) shows the 2-triangle sphere. Note that 0 is at the lower vertex, ∞ is at the upper; the midline is the negative real axis, and 1 is simultaneously at the left and right vertices of the kite; indeed, the positive real axis lies along both the left and right edges, points at the same level being identified.

Figure 7.4(b) shows our lifted triangulation, 6 triangles of each type forming a topological sphere. (As usual in our schematics, •, ×, ∘ mark vertices associated with 0, 1, and ∞, respectively.) The associated Belyĭ map $R : \mathbb{P} \longrightarrow \mathbb{P}$ is determined up to precomposition with a Möbius transformation of $\mathbb{P}$; this allows us to normalize so that $R(0) = 0$, $R(1) = \infty$, and $R(\infty) = \infty$, as indicated in the schematic. The combinatorial symmetry of the triangulation about the real axis in the schematic implies by Schwarz reflection a corresponding conformal symmetry. In other words, R will be real on the real axis and each half plane in its domain will be a union of 6 lifted triangles.

This last observation and the fact that we normalized so $R(1) = \infty$ are crucial, for they mean that when we iterate lifting under R we can keep track of the combinatorics. Figure 7.4(b) is, of course, the first lift of (a) under R; a second lift results in the schematic (c). Note that each triangle of Figure 7.4(b) decomposes into 6 smaller triangles in (c), each of which is mapped conformally under $R^{(2)} = R \circ R$ to the corresponding hemisphere (shaded or unshaded) of (a). In a similar fashion, one can lift under an arbitrary number of backward iterates of R, that is, iterates of R^{-1}.

Now, where are the "pentagons"? They're formed at each lifting stage by the points mapping to the real interval $[1, \infty)$. Observe in Figure 7.4(b), for example, that R has a branch point of order 4 at a certain point on the midline; this point corresponds by symmetry with some negative real number a for which $R(a) = 0$. That branching accounts for the "pinwheel" of 10 triangles meeting at a in (b): namely, they are the lifts of the two triangles meeting at 0 in (a). The outer edges of the triangles forming this pinwheel get mapped to $[1, \infty)$ under R, and they form one pentagon. Lifting a second time to get (c), that pinwheel is replicated 6 times at the 6 lifts of a — that is, at the 6 non-zero lifts of 0 under $R^{(2)}$, each of order 4. In Figure 7.4(d) we have highlighted $R^{(-2)}[1, \infty)$ to make the pentagons more visually evident. Note that the original pinwheel of (b) decomposes into 6 pinwheels; that is, the lift under R implements the pentagonal subdivision rule of Figure 7.2.

Of course these images are just schematics, but when the conformal structure is imposed the schematic pinwheels show that any pair of conformal pentagons sharing an edge are also conformal reflections of one another. So these are conformally correct regular pentagons.

Recalling that the left and right edges of a schematic are to be identified, there are at each lifting stage only two triangles at the origin — in other words, the origin doesn't have a complete pinwheel. To fix this, we take a fifth root, meaning that we take one further preimage under the auxiliary function $g(z) = z^5$. This replicates our schematic 5 times about the origin as in Figure 7.5. Since g^{-1} is conformal away from the origin (and ∞), the preimages of our pentagons retain their conformal and reflective properties.

Summarizing, for each $n \geq 1$ one obtains the nth stage decomposition D_n of the sphere into 6^n conformally regular pentagons by: (1) lifting the interval $[1, \infty)$ to $\mathbb{P}$ under $R^{(n)}$; (2) lifting that result to $\mathbb{P}$ under $g(z) = z^5$. Each pattern D_{n+1} is obtained from D_n by breaking each of its pentagons into 6 conformally regular pentagons according to the subdivision rule.

We are almost there — only a renormalization is missing. If p_0^n denotes the "base" pentagon in D_n, the one centered at the origin, then with each subdivision that base pentagon is reduced in size (and rotated by $2\pi/5$). Intuition suggests that rescaling by an appropriate complex number λ will stabilize this base pentagon. Putting the fifth root aside for a moment, we are contending with iterates of $h = R^{-1}$ in a neighborhood Ω of the simple fixed point 0, so at least asymptotically the scaling factor would have to be $h'(0)$. This is in fact a familiar situation in

66 7. ADDITIONAL CONSTRUCTIONS

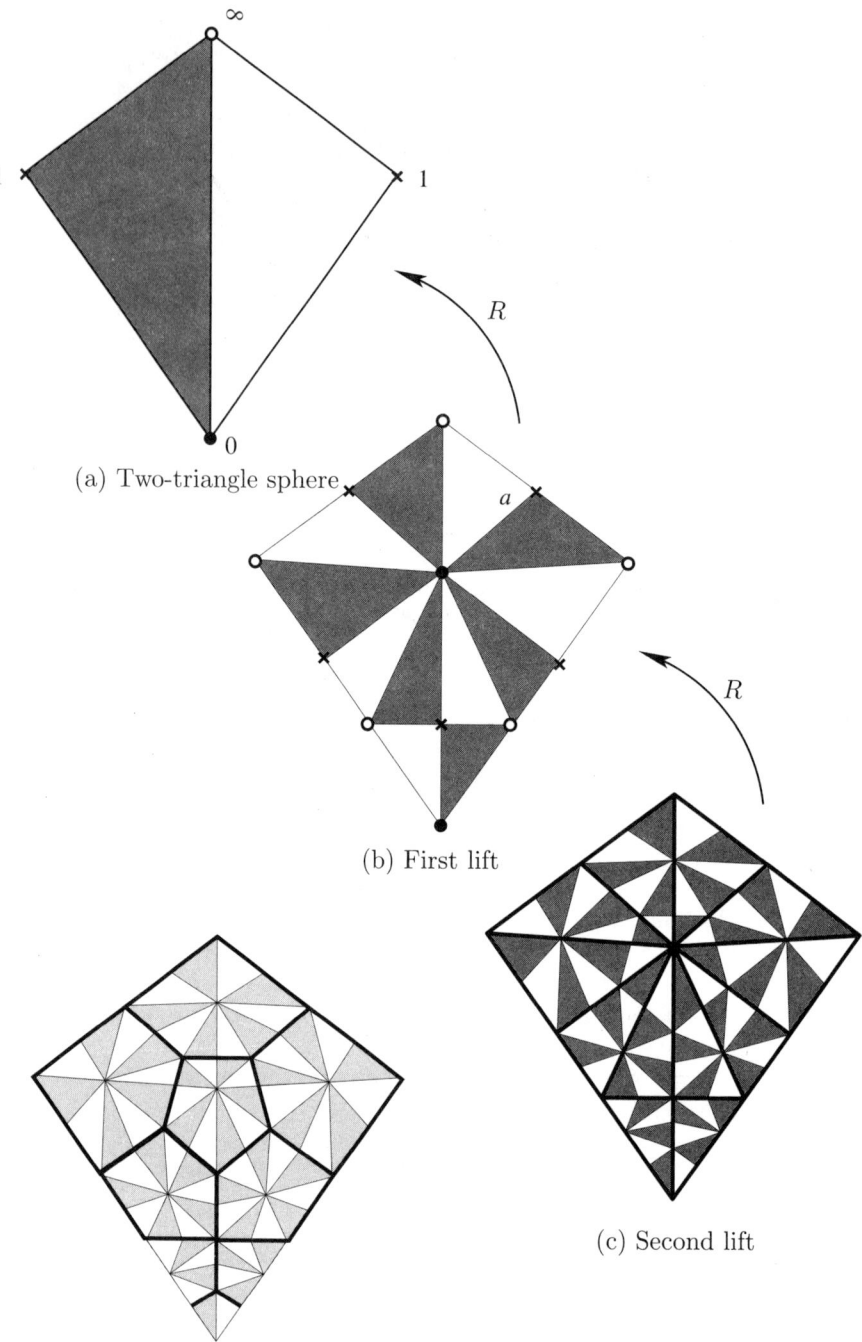

(a) Two-triangle sphere

(b) First lift

(c) Second lift

(d) Highlighted pentagons

FIGURE 7.4. Schematics of R^{-1} iterates

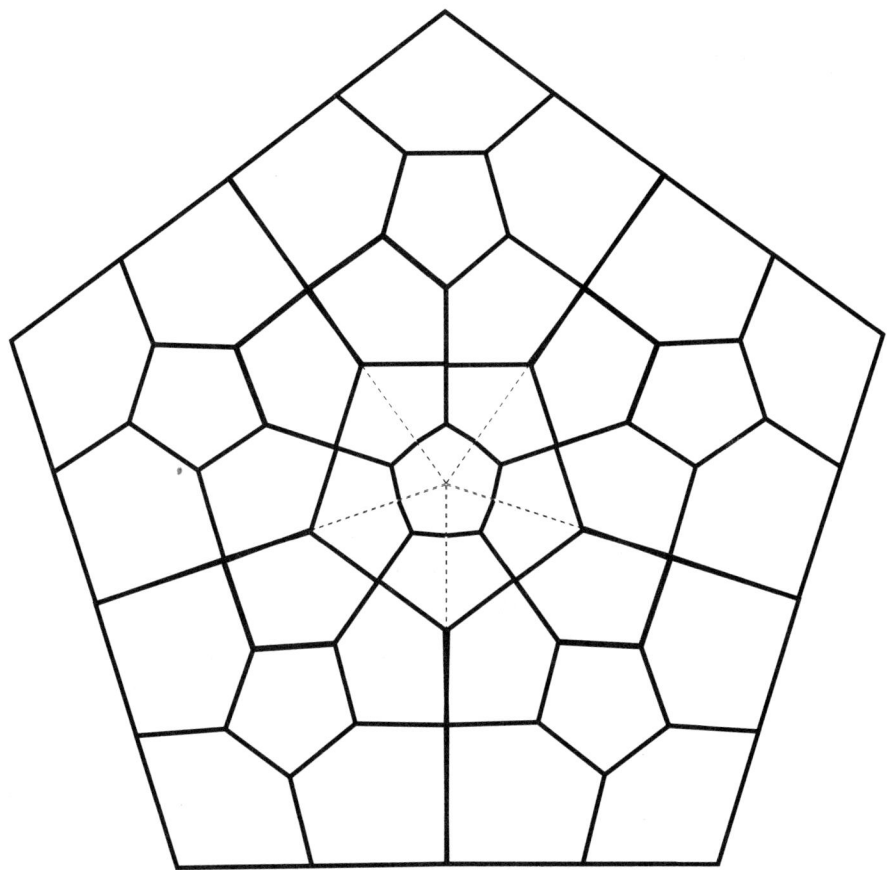

FIGURE 7.5. Fifth root of the schematic.

iteration theory, and the result we use is due to Königs, [**31**] (see [**46**, Chp. 6]). In particular, defining

(1) $$\phi_n(z) = h^{(n)}(z)/\alpha^n, \; z \in \Omega,$$

one finds that $\phi_n \longrightarrow \phi$ on Ω, where ϕ, appropriately called the **Königs function** for h, is a nonconstant function analytic in Ω and satisfying

(2) $$\alpha\phi(h(z)) = \phi(z), z \in \Omega.$$

In essence, this says that ϕ's behavior at the origin reflects the *infinitesimal* behavior of $h = R^{-1}$ there. Analytic continuation arguments show that ϕ is meromorphic on the complex plane and we can reintroduce the fifth power by defining

$$f(z) = \phi(z^5).$$

This is meromorphic on the plane and one finds that $f^{-1}([1,\infty))$ is precisely the *regular pentagonal* tiling represented in Figure 7.1. In particular, unwinding (1) and (2) shows that the scaling factor λ for the tiling is $\alpha^{1/5}$.

The final step, then, is to determine α. We make a few observations about R. Counting triangles shows R to be a 6-degree rational map, hence, by Riemann-Hurwitz, R has 10 branch points, counting multiplicities. Observe the following from Figure 7.4(b):

(1) $R(0) = 0, R(\infty) = \infty$ and R is locally univalent at both 0 and ∞.
(2) $R(1) = \infty$ with branching of order 1.
(3) $R(a) = 0$ at a point $a < 0$ with branching of order 4.
(4) $R(b) = \infty$ for some $e, 0 < e < d$, with branching of order 2.

These conditions were sufficient for Floyd, with the help of symbolic software, to determine R uniquely; in particular:

$$R(z) = \frac{2z(z+9/16)^5}{27(z-3/128)^3(z-1)^2} \text{ and } R'(0) = -324.$$

We conclude that our scaling factor λ is the algebraic number $\lambda = (-324)^{1/5}$, as anticipated. Of course, the fifth root of negative one accounts for the rotation by $\pi/5$.

Let's not forget that R is a Belyĭ map. From Figure 7.4(b) one obtains the dessin shown in Figure 7.6 with a refined circle packing approximating its conformally correct embedding.

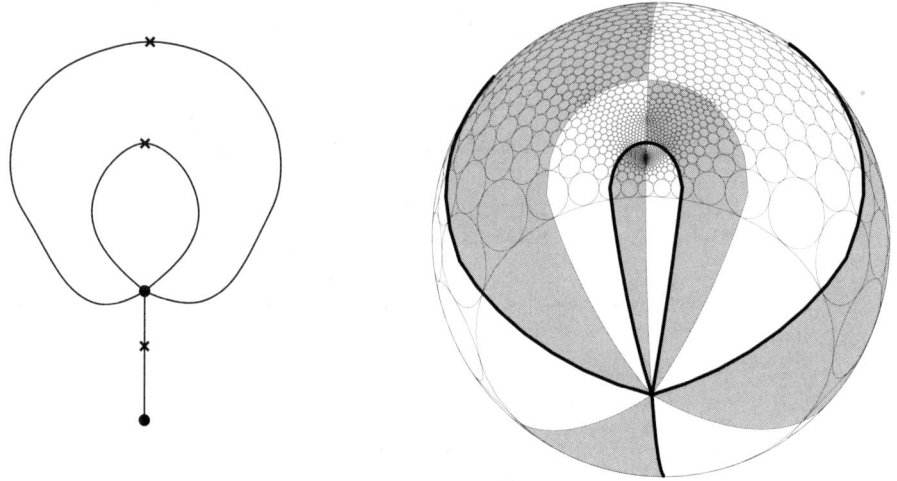

FIGURE 7.6. Dessin of the regular pentagonal tiling.

Before we leave tilings we should emphasize that they enjoy far greater variety than this example suggests. The methods of this monograph apply if, for example, one always proceeds by breaking the polygons into triangles and imposing the equilateral structure. In general, however, polygons lack the "three-ness" that makes things work so effortlessly with triangles; whereas any two euclidean triangles are conformally equivalent (vertices mapping to vertices being always understood),

7.2. The j function

Any triangulation T of a connected oriented topological surface S induces a conformal structure on S by treating the faces as unit equilateral euclidean triangles as in the dessin setting, though in general S would be open and/or bordered if T is infinite and/or has border. A Belyĭ-type meromorphic function f can always be constructed on S as before, using T if its faces can be alternately shaded or by applying a preliminary barycentric subdivision and using $T' = \beta T$ otherwise.

Such triangulations can arise, for example, as liftings of simpler triangulations under branched or covering maps, and that is the situation we will illustrate with the *j-function* $j(z)$ of classical function theory (see, for example, [**1**]). Our approach, as with dessin, is to define our objects purely combinatorially, to induce the equilateral structure, and then to look for algebraic properties of the resulting functions — of course we know that striking algebraic properties emerge here.

The real axis divides the sphere into two hemispheres, each to be treated (as before) as a triangle with vertex set $\{0, 1, \infty\}$, the lower half plane being shaded. Consider an abstract unlimited branched covering ϕ of the punctured sphere $\mathbb{P}\backslash\{\infty\}$ with the property that ϕ branches only over 0 and 1, with simple branching (order 1) at every 1-point and branching of order 2 at every 0-point. (Infinity is termed a *logarithmic* branch point in this setting.) Our 2-triangle triangulation of the sphere lifts under ϕ to an infinite, alternately shaded, abstract triangulation T of a topological disc D. Each vertex of T mapping to 0 (resp. 1) is shared by 6 (resp. 4) faces. The equilateral surface structured on T is conformally equivalent to the upper half plane H (the hyperbolic plane), and if the *in situ* triangulation T is appropriately normalized, then the resulting analytic function from H to the punctured sphere is $j(z)$.

The conformally correct triangulation of H is shown in Figure 7.7. The algebraic properties of $j(z)$ are well known and much studied, see [**1**] for a survey. The darker lines in the figure highlight a pair of faces defining the standard fundamental domain for the action of $PSL(2, \mathbb{Z})$ on H; this region is the moduli space for conformal 1-tori, while H is the Teichmüller space. In particular, then, our abstractly conceived triangulation has resulted in a function $j(z)$ which is the basic invariant for $PSL(2, \mathbb{Z})$.

Having seen this construction of the j-function, it's natural to specify branch orders other than $(1, 2)$; Figure 7.8(a) and (b) illustrate conformal triangulations for values $(2, 3)$ and $(4, 5)$ in the unit disc rather than the half plane. These are hardly new: given orders $m \geq 1$ and $n \geq 1$ and ∞ (at the puncture), each conformal triangle in H is actually a hyperbolic triangle having angles $\frac{\pi}{p}, \frac{\pi}{q}$, and ∞, where $p = m + 1, q = n + 1$, and the triangulation can be obtained by repeated Schwarz reflection in its edges. Locally this is just the reflective structure we exploited earlier, but here the global nature of the reflections accounts for their algebraic

FIGURE 7.7. The j-function

link. These reflections comprise the classical "triangle group" $\langle \frac{1}{p}, \frac{1}{q}, 0 \rangle$, but other triangle groups fit equally well with circle packing. For the classical background, see [**36, 27**]. The Klein surface of Example 10 is associated with the $\langle 2, 3, 7 \rangle$ triangle group in the hyperbolic plane, while the next section exploits triangle groups on the sphere.

The lesson to take from these examples is that one could do much more general infinite triangulations: their combinatorics alone determine conformal structure and circle packing can serve to approximate the result. We've illustrated here that internal symmetries can both simplify the construction (since refinements are unnecessary) and lead to associated algebraic structure. In the next example we are forced to rely on such internal symmetries because circle packing theory on the sphere remains incomplete.

7.3. Schwarz Triangles

Meromorphic functions on compact surfaces are always combinatoric in a certain sense due to their branch structures. However, we have enjoyed (at least) two

7.3. SCHWARZ TRIANGLES

FIGURE 7.8. Functions for other triangle groups.

special advantages in the examples we have constructed, the Belyĭ maps and the j-function. First, information in the range has always been lifted back to the domain; namely, two triangles forming the range sphere are lifted, using a prescribed branch structure, to form a triangulation of the domain. Second, any two lifted triangles are conformally equivalent, meaning that a combinatorial structure is tantamount to a conformal structure. In approaching more general meromorphic functions one cannot so easily finesse the conformal rigidity issues.

We need first to comment on branching as manifest in our circle packings so far, since the lifting in our constructions tends to obscure what is going on. Figure 7.9 illustrates a typical simple branch point for a discrete, hex refined, Belyĭ map b_D. Four faces meet at the darkly shaded circle c in the domain, which consequently has eight petal circles in its flower. Two faces meet at its image circle $\tilde{c} = b_D(c)$ in the range, with it's four petal circles. Of course in the mapping the four faces at c wrap twice around $\tilde{c}$. It is common in classical function theory, however, to picture an "image surface" for the mapping which *projects* to the range. In other words, in Figure 7.9(b) one can treat $\tilde{c}$ as having a flower with eight petal circles which happen to align in pairs.

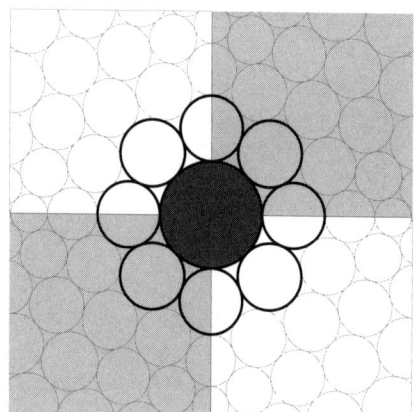 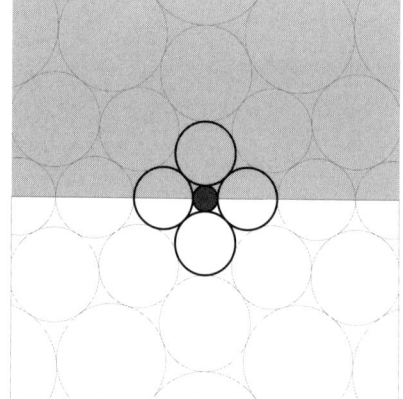

FIGURE 7.9. Behavior at a simple branch point.

If we suppose b_D is a k-to-1 map and apply this same thinking globally, then the image surface of b_D is pictured as a k-sheeted covering surface of the sphere — k distinct circles from the image surface project to each circle of the range packing. In more general "non-lifting" constructions, one is faced with the task of creating a circle packing for the image surface without the advantage of this ubiquitous alignment. One would like the capability, for example, to specify branch *points* (in the domain) rather than branch *values* (in the range). In the discrete theory, that translates into starting with a circle packing in the domain, specifying the circles where branching is to occur, and then creating the branched image packing on the sphere.

Unfortunately, the sphere is a difficult circle packing environment; neither the existence and uniqueness results nor a working packing algorithm for branched packings are known. The special case of "polynomial" branch structures can be handled by moving to the hyperbolic plane. However, the best we can do here is to illustrate the general goal with one pretty but isolated example which fell into our laps when contemplating reflective triangulations ([**7**]).

Reflective triangulations of the sphere were classified by Schwarz [**17**]. All are generated by the basic $\langle 2, 3, 5 \rangle$-*Schwarz triangle*, that is, a triangle having angles $\pi/2, \pi/3, \pi/5$. An example, denoted t, is illustrated in Figure 7.10(a). (For later reference, it is also equipped with two circular arcs.) Repeated reflections of an initial Schwarz triangle in its sides will generate a tiling of $\mathbb{P}$ by 120 Schwarz triangles as in Figure 7.11.

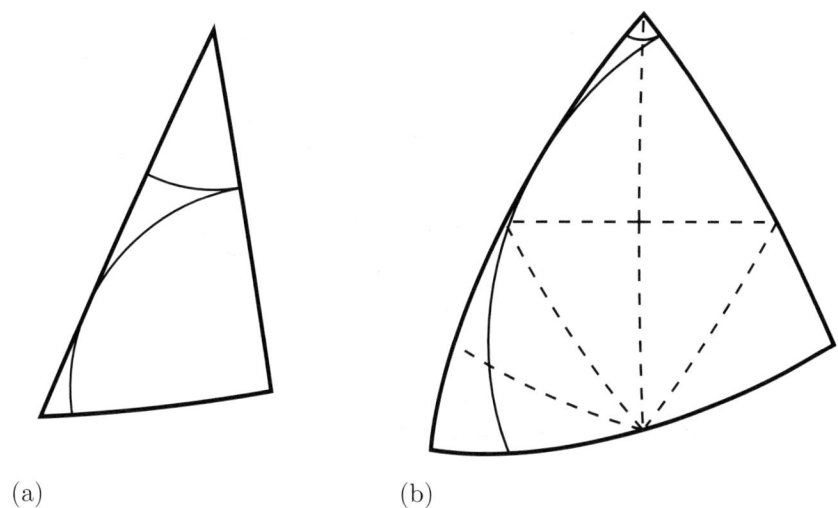

(a)   (b)

FIGURE 7.10. Schwarz triangles t and T.

Grouping seven Schwarz triangles together as in Figure 7.10(b) creates what is called a $\langle 2, 3, 5/2 \rangle$-triangle, which we will denote by T. (Again, two circular arcs have been drawn on T.) Place a copy of T on the sphere; repeated reflections in its edges will also close up on the sphere, again forming a pattern of 120 of these larger triangles. Simple branching occurs at each vertex of angle $2\pi/5$, for when orientation is taken into account, it requires 10 reflections about such a vertex to close up, wrapping twice around for a total angle of 4π. There are 12 points where this occurs, distributed as the 12 vertices of a regular dodecahedron. The euler characteristic and Riemann-Hurwitz formula imply that the generic point of the sphere is covered by seven of these larger triangles — consistent with 120 copies of T, each 7 times as large as t.

The conformal map from t to T of Figure 7.10 generates, by repeated Schwarz reflection, a degree seven meromorphic function of $\mathbb{P}$ onto itself, i.e., a degree seven

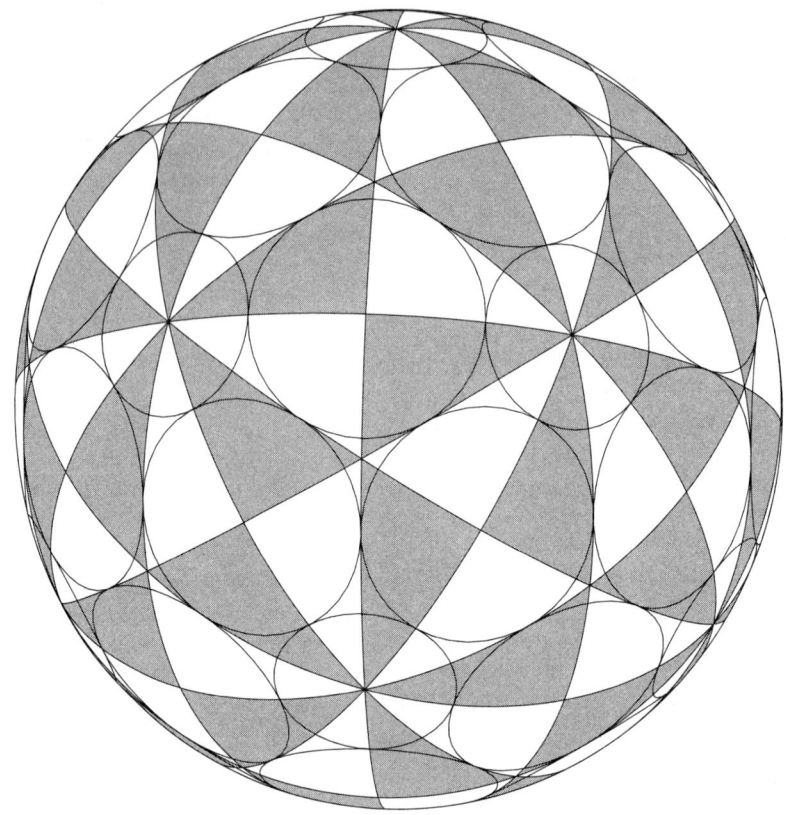

FIGURE 7.11. 120 Schwarz triangles tiling the sphere.

rational map. Appropriately normalized, this is precisely the function

$$F(z) = \frac{z^2(3z^5 - 1)}{(z^5 + 3)}.$$

Now consider the circular arcs we drew on t and T. In each case, carrying these along in the constructions results in a circle packing. The packing $\mathcal{P}$ associated with t can be seen in Figure 7.11. The packing $\widetilde{\mathcal{P}}$ associated with T is shown in Figure 7.12(a); the branch circles are shaded. This is rather difficult to get a handle on, so in Figure 7.12(b) we show a single branched flower, the central shaded circle being encircled twice by the chain of five successively tangent neighboring circles.

The discrete rational function f defined by mapping $\mathcal{P}$ to $\widetilde{\mathcal{P}}$ is the analogue of the classical map F. In fact, due to the pervasive symmetries involved here, after appropriate normalization, f will precisely interpolate F at its 12 branch points, indeed at all the circle centers.

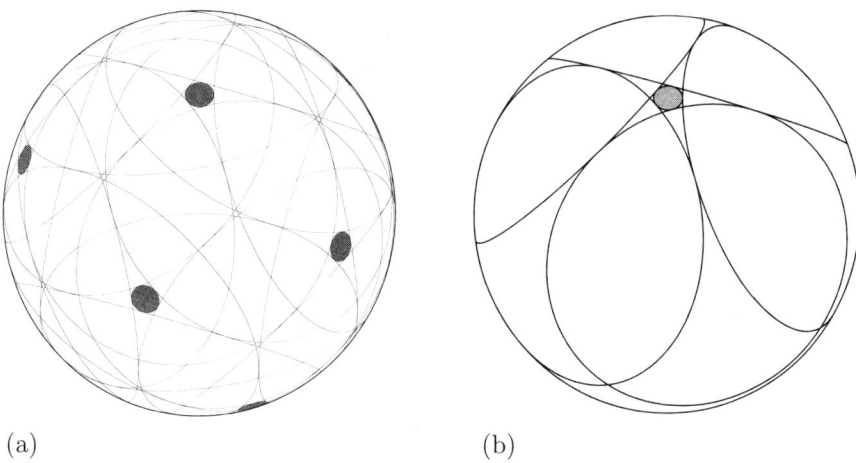

(a)   (b)

FIGURE 7.12. The branched packing $\widetilde{\mathcal{P}}$ and one of its 12 branched flowers.

7.4. Graph Embedding

The carrier of a univalent circle packing in a surface provides an embedding of its complex $\mathcal{K}$ (in fact, in the cases we have been considering, it even provides the geometry for the surface). In some cases it may not be the surface which is of interest, but rather the embedding of the 1-skeleton $\mathcal{K}^{(1)}$ as a graph. Indeed, circle packings are valuable for embedding fairly general locally planar graphs: the 1-skeletons of their carriers are geodesic, they can simultaneously provide geodesic dual orthogonal graphs, and graph edges meet at angles bounded below by constants depending only on degree. Graphs of considerable complexity can be generated automatically and randomly and the packings computed relatively quickly; packing algorithms are known that work in polynomial time with the number of vertices.

The circle packing techniques of the paper provide embeddings of general triangulated surfaces, polygonal surfaces, open and bordered surfaces, and are being applied in recent work on conformal "tilings" (see [**8**]). Circle packings tend to be accompanied in all these settings by vestiges of classical analysis — random walks, discrete Laplacians, discrete extremal length, discrete analytic functions, and so forth.

Some of the interest in dessins d'enfants stems from fact that they provide a countably dense set of points of Teichmüller space (see the next section), and that these points follow an underlying organizational structure. Physicists, for instance, are interested in statistical and asymptotic analysis of surfaces.

CHAPTER 8

Non-equilateral Triangulations

A triangulated surface S may come equipped with a piecewise affine structure in which the triangles are not necessarily equilateral. That structure still determines a conformal structure, raising the issue of how to carry out conformal approximations. In this chapter we briefly describe two methods with the potential to addressing such cases: "discrete conformal welding" and "inversive distance packings". We advise the reader that open questions remain in each approach, so our comments here are preliminary in nature.

The first is an elementary application of much more general discrete "conformal welding" procedures under development by Williams [53]; the open questions here concern practical implementation. The second involves "inversive distances" and extends basic notions of circle packing; as it turns out, the computations seem to work in practice, but the theory can't yet account for the behavior.

8.1. Welding Approach

Circle packing maps are quasiconformal with a bound on dilatation κ depending on degree. In regions of S where the packings become more deeply embedded in purely hexagonal generations during refinement, the local dilatation approaches 1 and the maps become progressively more conformal. The fact that dilatations bounded away from 1 persist at the vertices of S does not effect global convergence because that distortion is restricted to sets whose areas, in the piecewise euclidean metric of S, go to zero. This fact was key to our work in Chapter 4. In a similar way, if that distortion is restricted to neighborhoods of the edges of S having areas which go to zero under refinement, the global limit is again conformal. This gives us an amazing amount of freedom in the transitional combinatorics between triangular faces.

In general terms, this is a "conformal welding" matter. Suppose one is given surfaces S_1 and S_2 endowed with conformal structures and having border segments B_1 and B_2, respectively, and suppose that an orientation reversing homeomorphism $\phi : B_1 \longrightarrow B_2$ is specified. One can attach S_1 and S_2 to obtain a surface S by identifying $x \in B_1$ with $\phi(x) \in B_2$; if there is a conformal structure on S which is compatible with those of S_1 and S_2, then we say that ϕ is a *conformal welding homeomorphism*. Using the Riemann Mapping Theorem one can assume without loss of generality that B_1 and B_2 are segments of the real axis. Various conditions on ϕ will ensure that it is a conformal welding map: bilipschitz and quasisymmetric are prime examples in the extensive literature [34].

In these classical terms our situation is dead simple: S_1 and S_2 are euclidean triangles, B_1 and B_2 are edges, and $\phi : B_1 \longrightarrow B_2$ is affine. Conformal welding consists of scaling so B_1 and B_2 have the same euclidean length and then simply butting the triangles edge-to-edge in the plane. The discrete version, however, has its sublties; though we enjoy great latitude in choosing our combinatorics, we ultimately have to make the process consistent with refinement in order to ensure that the limiting welding map is actually ϕ. Discrete conformal welding is being developed by Brock Williams; we refer the reader to [53, 54, 55] for details.

We content ourselves with an example, mimicking the classical welding of two specific euclidean triangles shown in Figure 8.1(a). Start by nearly filling each triangle with a portion of a regular hexagonal packing and abutting them as in Figure 8.1(b); scaling has given the circles different sizes. Now add extra vertices and edges as indicated by the dashed lines in (c), forming *in situ* a single simply connected "welded" triangulation. This new triangulation is circle packed in (d); the attachment edges are presented as thick lines. The process can be repeated in refinement stages, but each requires it own pasting decisions. As shown by Williams, the only requirement in adding the attaching vertices and edges (as in Figure 8.1(c)) involves control of the angles *in situ* of the triangles they form. If these angles are bounded below by some $b > 0$ at all refinement stages, the embedded triangles provided by the circle packings will converge to the classical pasting of Figure 8.1(a). (Note that we have chosen our example so the *outer* edges of the triangles lie along axes in their hexagonal packings. This allows us to artificially maintain the angles of the quadrilateral region so we can compare the results; in practice, these other edges would be affected by pastings of their own.)

8.2. Inversive Distance Packings

A second approach to nonequilateral triangulations is cleaner and more easily implemented; this approach relies on *inversive distance packings*. There is both good news and bad news here. The bad news is that the theoretical underpinnings are not yet in place, despite the fact that the open issues are straightforward generalizations of the tangency case. The good news is that it appears to work fine in practice with only minor adjustments. In any case, what we discuss in this section is speculative until the theory advances or the usefulness overcomes its shortcomings.

The key is that the relative positions of two circles c_1, c_2 in the plane can be described in a Möbius invariant way by a continuous parameter $t = t(c_1, c_2) \in [0, \infty)$. Fix c_1, say, and move c_2 from infinity towards c_1. While the circles remain disjoint, t is the classical inversive distance between c_1 and c_2 and it decreases as the separation decreases. As c_2 reaches tangency with c_1, t reaches 1. Continuing to move c_2 towards c_1, the circles begin to overlap with overlap angle θ, and then t is defined as $\cos(\theta) \leq 1$. We will stop when t reaches 0, at which point c_1 and c_2 intersect at right angles. We will refer t as the inversive distance parameter, and for use here we'll restrict to $t \in [0, \infty)$. If ϕ is a Möbius transformation of the plane then $t(\phi(c_1), \phi(c_2)) = t(c_1, c_2)$. (Note that Rivin [26] considers circle packings with parameters $t \in [-1, 1]$; we stick with $t \geq 0$ because fundamental geometric and computational difficulties arise when $t < 0$ is allowed.)

An **inversive distance** circle packing $\mathcal{P}$ is a configuration of circles with a specified pattern of inversive distances. In particular, there is an underlying complex $\mathcal{K}$, as usual, but also a map τ from the edges of $\mathcal{K}$ into $[0, \infty)$. A circle packing for $(\mathcal{K}, \tau)$ is defined as before except that one replaces the tangency requirement, $t(C_u, C_v) = 1$, by the requirement $t(C_u, C_v) = \tau(e)$ whenever $\langle uv \rangle = e$ is an edge of $\mathcal{K}$. Figure 8.2 shows two inversive distance packings for the same complex $\mathcal{K}$ (that of the coarse packing in Figure 7.3); (a) has randomly assigned overlaps (inversive

8.2. INVERSIVE DISTANCE PACKINGS

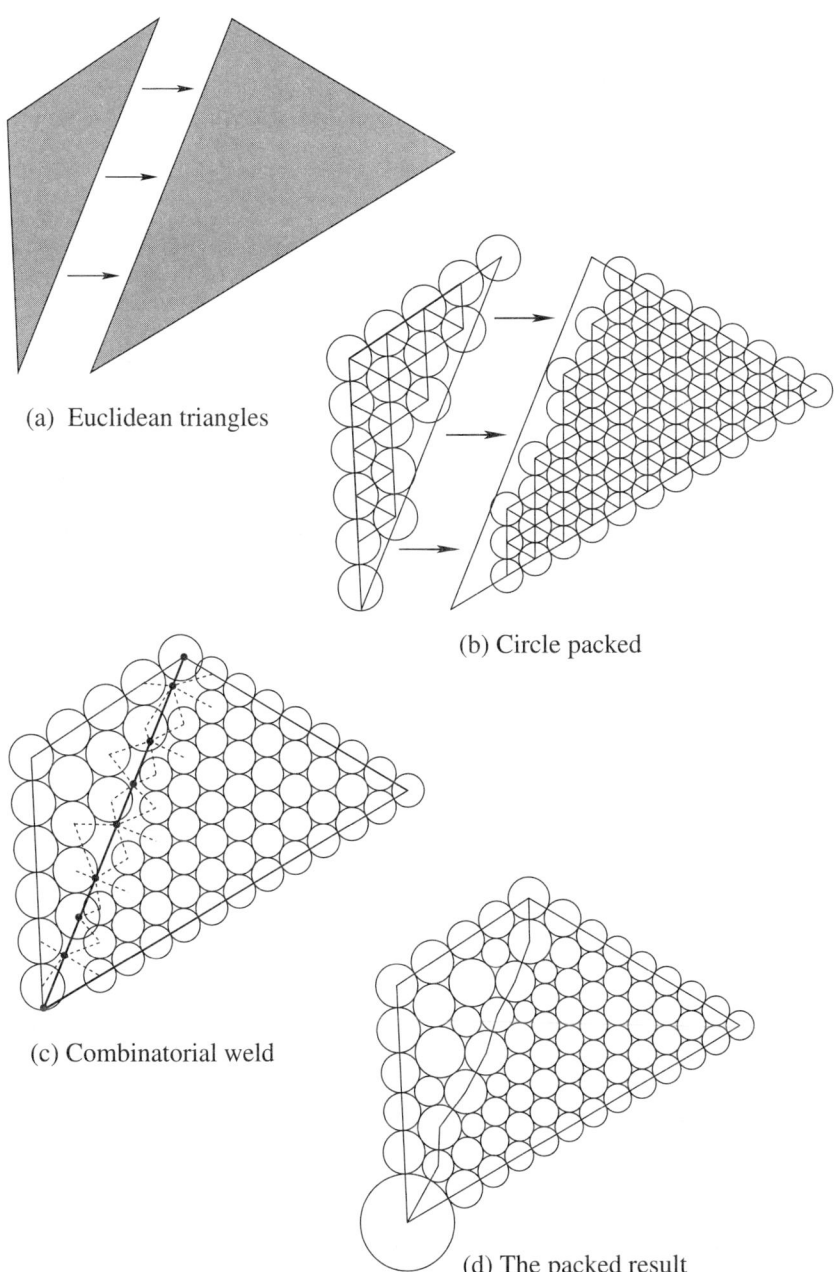

FIGURE 8.1. Discrete conformal welding

distances $0 \le t \le 1$), (b) has randomly assigned separations (inversive distances $1 \le t \le 5$).

The original Andreev-Thurston result for circle packings of the sphere was proven for overlapping circles with angles of overlap $\theta \in [0, \pi/2]$, in our terminology,

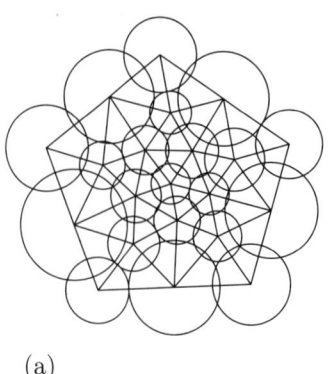 

(a)   (b)

FIGURE 8.2. Inversive distance packings: random overlaps, random separations

inversive distance packings with parameters $\tau(e) \in [0,1]$ for all edges e. (There are also some minor combinatoric side conditions which need not concern us here; see [**39**].) Z-X. He [**24**] has proven uniqueness for infinite packings of the plane under this same restriction $\tau(e) \in [0,1]$. However, as you will see, our interest lies with inversive distance packings with parameters $t \geq 1$; we refer to these as **S-packings** (S for "separated").

Let's explore briefly how S-packings could be used — we'll come back to the theoretical issues momentarily. Figure 8.3(a) illustrates a topological closed disc Ω triangulated by euclidean triangles. Our aim is to approximate the conformal mapping of Ω onto a euclidean rectangle, requiring that designated vertices be mapped to the corners. The example we've chosen is planar simply for convenience; it could as well be a fragment of a surface in 3-space with euclidean triangles as faces.

Write $\mathcal{K}$ for the abstract complex represented by the given triangulation. Create an S-packing for $\mathcal{K}$ as follows: Let m be the minimum edge length in the triangulation of Ω, center a circle of euclidean radius r_0 at each vertex of the triangulation, record for each edge e of $\mathcal{K}$ the inversive distance $\tau(e)$ for the circles at its ends. A fortiori the circles form an S-packing $\mathcal{P}_0$ for $(\mathcal{K}, \tau)$, pictured in Figure 8.3(b). Prescribing appropriate boundary angle sums and repacking gives the packing $\mathcal{Q}_0$ for $(\mathcal{K}, \tau)$ of Figure 8.3(c). The map from $\mathcal{P}_0$ to $\mathcal{Q}_0$ is morally a discrete conformal map, though we don't yet have the right to use that term. Justification would be provided if we could refine the packings and establish convergence of the resulting maps to a classical conformal mapping.

At least the mechanism of refinement, if not the theory, is in excellent shape with this approach. Hex refine the original triangulation *in situ* by placing a new vertex at the euclidean midpoint of each edge. The resulting refined complex is the familiar $\mathcal{K}^{(1)}$, but the geometry is especially kind to us: each of the original triangles is broken into four which are similar to it with scale factor $1/2$. Setting a common radius $r_1 = r_0/2$ at all vertices, new and old, observe that each new triangle inherits

its three inversive distances from its parent, so the edge parameters $\tau^{(n)}$ are easily replicated. In other words, forming $(\mathcal{K}^{(n)}, \tau^{(n)})$ is computationally trivial. Indeed, as a practical matter you can experiment now: CirclePack accommodates arbitrary inversive distance assignments τ, implements hex refinement, and typically computes and displays resulting circle packings without complaint. Figure 8.3(d), (e) show S-packings for $n = 1$ and $n = 2$ packed as rectangles. From our experience in the tangency setting, it's difficult to avoid jumping to the conclusion that the limit of the discrete mappings suggested by Figure 8.3(f) is conformal.

The pieces of theory needed to back up our speculation are familiar from the tangency case. (•) First is the existence of S-packings under various boundary conditions. Inconsistencies are now possible among the combinatorics, radii, and inversive distances, even for simple flowers. Some type of moduli space of consistent configurations is needed. Nonetheless, we should point out that in practical situations, with a closed triangulated surface and a known S-packing as base, inconsistencies don't seem to emerge. (•) The next issue is uniqueness, but that seems to be bound up with existence; there is an appropriate version of the Ring Lemma and key notions of monotonicity continue to hold, but the Perron arguments depend on regimes which avoid incompatibilities. (•) The final piece, convergence, pivots on uniqueness of infinite S-packings. Under refinement, a point interior to an original parent triangle is surrounded by ever more generations of purely hexagonal combinatorics, all inheriting an identical pattern of the three inversive distance parameters from the parent triangle. Conformality in the limit results from uniqueness of the corresponding infinite hexagonal packing of the plane. That seems likely to follow from probabilistic methods which have worked with tangency [**48, 19**] and overlapping packings [**24**]. So there is a lot of promise and a lot of work, a good situation all around.

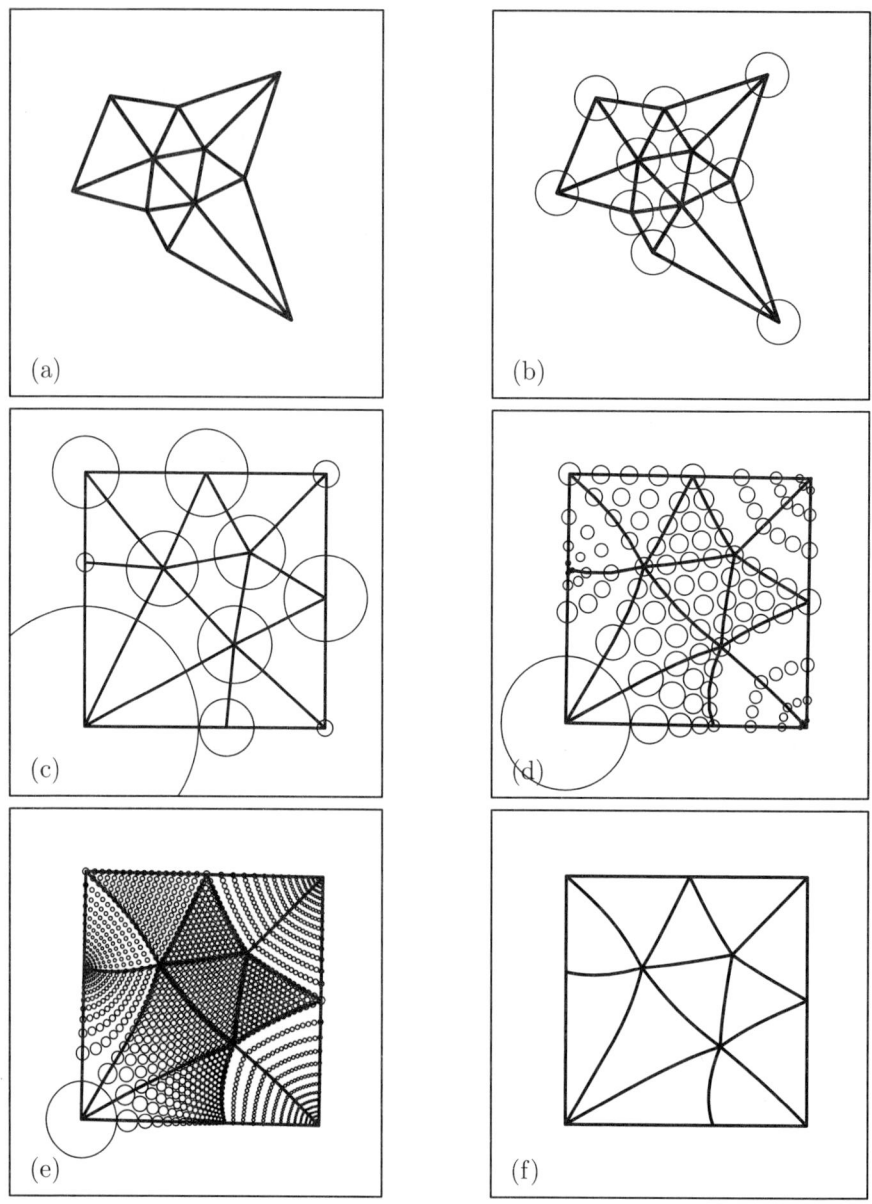

FIGURE 8.3. Examples of S-packings and refinements.

CHAPTER 9

The Discrete Option

We hope that this monograph has displayed the remarkable knack that circle packing theories have for both mimicking and approximating their continuous models in conformal geometry and analytic function theory. It might be said that they provide a "quantum theory which is classical in the limit". In this closing chapter we wish to propose a broad and somewhat independent view of the discrete theory, emphasizing its attractions and intrinsic mathematical issues.

To set the stage, however, let's first discuss *approximating* the classical theory, which for some will clearly remain as the main focus. We have demonstrated several contributions that the discrete setting can make: visualization, manipulation with procedures such as "dessin modification", and rough numerical estimation, to mention a few. The main strength of the discrete theory seems to lie in the experimental setting which it provides for developing intuition and perhaps new theoretical insights. It is not yet suitable for highly accurate numerical approximations, as are available, e.g., with genus zero trees in the dessin setting [**16**] or with classical numerical conformal mapping methods for plane domains, [**52**]. Moreover, there is considerable work to be done on what are considered the standard measures of accuracy, numerical precision, rates of convergence, complexity, and so forth. For example, there is a rapid rise in computation times for circle packings with successive refinement, and even with numerical improvements one might wonder about the long range practicality for actual estimation. This is somewhat counterbalanced, however, by the surprising accuracy of even coarse packings; this accuracy is one of the more intriguing issues *vis-a-vis* the classical theory. Questions such as the Teichmüller distance from a dessin surface S_D to its discrete analogue s_D are of considerable interest, but probably very hard to pin down. In spite of current numerical limitations, however, we should emphasize that circle packing is the only game in town for approximating many general conformal structures. It has certainly shown enough prowess to warrant continued numerical development.

We want to leave these *quantitative* issues now and try to move our thinking wholly to the discrete setting and the *qualitative* behavior there. To this end, *suppose you were resigned to living entirely in the **discrete world***. What would you see as the key *intrinsic* issues? What internal structures do the collections of surfaces and functions have? fail to have? How much algrebraic structure can one see? What can one learn when your computer can actually build surfaces? What classical parallels emerge and what wholly new questions arise?

9.1. Dessins

To kick off the discussion, let's return to the dessin setting and consider dessins of a fixed positive genus g as viewed from the standpoint of the Teichmüller space Teich(g). On the discrete side, we will restrict attention to the "coarse" dessin surfaces only — foregoing refinements. Belyĭ's theorem and the density of algebraic numbers in the complex field imply that the set of (classical) dessin surfaces

$$\mathrm{T}^*(g) = \{S_D : \mathrm{genus}(D) = g\} \subset \mathrm{Teich}(g)$$

form a countable dense subset. Indeed, these surfaces are characterized as those having defining equations with coefficients in a number field. This is one of their principal attractions; they provide a geometric playground for the study of algebraic numbers and Galois groups.

We introduced in Section 3.2 the parallel discrete (coarse) dessin surfaces, and we will denote this collection by

$$\mathrm{t}^*(g) = \{s_D : \mathrm{genus}(D) = g\} \subset \mathrm{Teich}(g).$$

This provides new playground equipment.

There are many natural questions about $\mathrm{t}^*(g)$. Note that it intersects $\mathrm{T}^*(g)$, as with the Klein surface (e.g., in the presence of sufficient symmetry), see Example 10. Are there any other intersections? Intuition suggests that these sets are essentially disjoint. Are there redundancies in $\mathrm{t}^*(g)$, in the absence of symmetry? Within $\mathrm{T}^*(g)$ there are systemic redundancies: for instance, the 1-skeleton of $\mathcal{T}(D)$, treated as a graph, leads to a new dessin D', but Proposition 4.1 implies that $S_D = S_{D'}$. Other redundancies are perhaps much deeper, as with the distinct dessins yielding the Picard surface of Example 11. (Incidently, with modifications of the proof of Theorem 4.1, one can prove that $\mathrm{t}^*(g)$ is a countable dense set in $\mathrm{Teich}(g)$; see also [**6**].)

A quite fascinating open question concerns the characterization of the points of $\mathrm{t}^*(g)$. Here there is a tantalizing new connection to algebraic numbers proven by McCaughan [**38**]: *The covering group for a point of* $\mathrm{t}^*(g)$ *is conjugate to a subgroup of* $\mathrm{PSL}(2, \overline{\mathbb{Q}})$ *(resp.* $\mathrm{PSL}(2, \overline{\mathbb{Q}} + i\overline{\mathbb{Q}})$*) if* $g > 1$ *(resp.* $g = 1$*).* That is, one may choose the covering group so that its matrix representatives have algebraic entries. For example, in the genus 1 cases, the modulus will be algebraic.

The ability in the discrete setting to experiment with quite complicated dessins suggests new issues. What, for example, are the algebraic implications of dessin modification? Are there geometric descriptions of dessin Galois orbits, or is this truely an algebraic notion? Can one actually do arithmetic with surfaces? In what circumstances can connections among tilings, rational iteration, Königs functions, and dessins be established as in §7.1. Just how suitable is this new playground equipment if you are interested in number fields and Galois theory?

Some of the interest in dessins d'enfants, in string theory, for instance, stems largely from the fact that dessins impose an "organizing principle" on the space of surfaces through the dense sets $\mathrm{T}^*(g)$. The discrete theory may offer definite advantages. Surfaces of considerable complexity can be generated automatically and randomly and the packings computed (at least at the coarse stage) relatively quickly, aiding in the statistical and asymptotic analysis of random surfaces. Circle packings seem to provide a natural geometry of the type that nature might favor, and their refinements are akin to the physicist's renormalization.

9.2. Function Theory

In Chapter 7 we generalized our dessin constructions to tiling patterns. We would like to demonstrate with another tiling how a discrete formulation and attendant experiments might suggest new techniques which then challenge the classical theory.

Our example involves the dodecahedral subdivision tiling created by Jim Cannon [9] and studied extensively by Cannon, Floyd, and Parry [12]. There are three tile "types" here: triangles, quadrilaterals, and pentagons. Figure 9.1 shows the rules for dividing each tile type into a prescribed pattern involving the same three types.

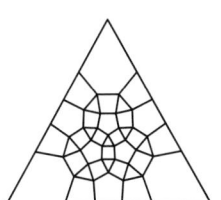  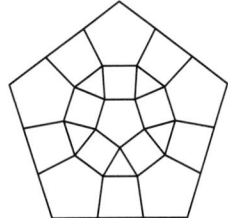

FIGURE 9.1. Dodecahedral subdivision rules.

Starting with a single quadrilateral tile and applying three stages of subdivision gives the pattern in Figure 9.2. Numerous geometric features of this beautiful picture might draw one's attention, but we'll be concentrating on the shape of the central rectangles.

Let's first discuss how the geometry arises. By starring each nontriangular tile at its barycenter, any pattern of tiles gives rise to a triangulation, and treating each triangle as a unit equilateral euclidean triangle then imposes a global conformal structure on the ensemble. (There are many alternative structures; for the conformal subtleties the reader is directed to [8].) As we have seen so often now, circle packings provide an accessible, faithful, discrete conformal model of this construction process, and when we speak about "conformal" shapes here, they are in fact discrete conformal shapes obtained from coarse circle packings. For example, our first image, Figure 9.2, is obtained by imposing the boundary conditions of a rectangle (representing the original quadrilateral tile), computing the circle packing, and then displaying only those edges in its carrier which are associated with the tile edges.

No one of these finite objects is our true target. As with the pentagonal tiling in Section 7.1, one can convert the subdivision rule into an expansion rule and then study the resulting infinite simply connected tiling pattern; the rules of Figure 9.1, interpreted as aggregation rules, describe how to form higher level tiles. The true issues are concerned with scale invariant features of the "spontaneous" global geometry. In particular, we will concentrate on the nested aggregate rectangular tiles at the center of the infinite pattern. Each has a conformal shape determined by that of the next higher level aggregate. Do these shapes have a nondegenerate asymptotic limit? An affirmative answer would suggest existence of a scaling like that occurring in the pentagonal case of §7.1 and all the attendant questions, such as whether the scaling factor is algebraic.

FIGURE 9.2. Stage 3 rectangle under the dodecahedral subdivision rule.

One sees only three levels of aggregation in the fragment of Figure 9.2, but already there is a trend in the shapes. At the very center is a single tiny quadrilateral tile (it is actually a conformal square by construction). The first aggregate rectangle containing it involves 31 tiles. The second aggregate is becoming more visible; by inspection it is clearly no longer a conformal square. A further aggregation brings in the whole of Figure 9.2; its rectangular shape is artificial, being imposed by our boundary conditions rather than by surrounding tiles. Unfortunately, building many more stages of aggregation is impractical: each stage increases the number of tiles by a factor of more than 30.

There's the rub: how to extract more detailed information? Living in the discrete world might suggest that we mimic the combinatorial self-similarity of aggregation. We'll call this **combinatorial conformal feedback** and illustrate with a rectangle R at the two-stage aggregation level.

Our working hypothesis is that in the (infinite) conformal tiling, the aggregate rectangle R will be a scaled copy of the smaller central rectangle r within it (shaded in Figure 9.3(a)). Proceed as follows: 1) Pack R as a rectangle, Figure 9.3(a). 2) Read off the ratio of radii on ∂r and feed these back to ∂R. More precisely, note that each vertex of ∂r is associated *via* the subdivision rule with a particular vertex of ∂R. Adjust the radii of ∂R so that the ratios among these circles match the ratios in ∂r; interpolate the radii of intervening vertices of ∂R. 3) Repack R with these new boundary radii, as in Figure 9.3(b). (4) Repeat steps 2 and 3 and watch the shapes stabilize. At this low refinement level the process stabilizes almost immediately to approximate the asymptotic similarity suggested in Figure 9.3(c): scaling by ≈ 8.2 carries ∂r onto ∂R. We are not going to settle this issue here, but our point is that there is no classical model for this conformal feedback.

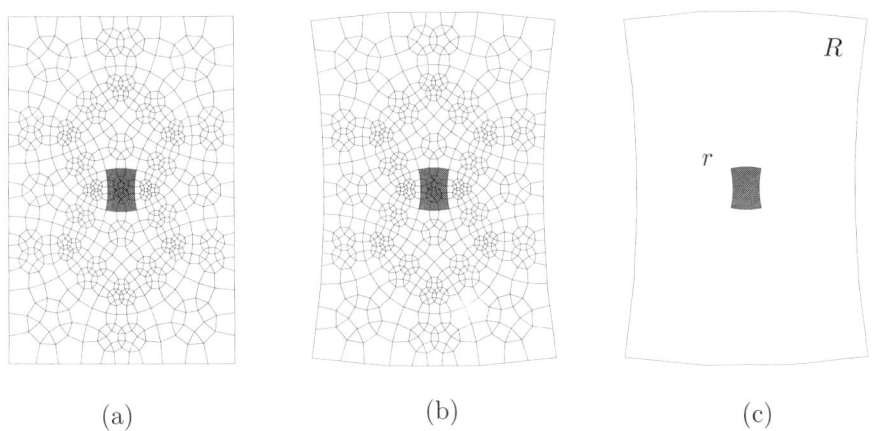

FIGURE 9.3. Combinatorial conformal feedback.

In closing, the authors suggest staying in this discrete world for a while longer — it has much of the geometry, combinatorics, and even possibly the number theory of the classical setting, it brings new ideas to the table, plus it has a certain charm all its own.

CHAPTER 10

Appendix: Implementation

The examples of the paper were produced using two suites of computer programs: `CirclePack`, a graphically based program for creating, computing, manipulating, and displaying circle packings, and `DesPack`, a set of auxiliary programs for building dessins and their associated data structures. These programs are available for use by others from the second author.

10.1. A Quick Experiment

As an overview, let us follow a typical development cycle for a dessin, say the genus 2 dessin of Example 7.

1. Start with the common fundamental domain for a genus 2 surface, schematically an octagon with "normal form" side-pairings. This is the minimal dessin of genus 2 and its canonical triangulation involves 16 triangles. These are numbered and the adjacency relationships are entered in an ASCII file (in a straightforward format) as input to `DesPack`. This file contains what we will informally refer to as a "red" chain, a chain of faces enclosing a fundamental region.

2. `DesPack` is programmed with the composite dessin moves of §6.1; three of these convert the basic dessin into the dessin D of Figure 5.10(a).

3. `DesPack` now generates a simple ASCII file representing a 2-complex $\mathcal{K}$ which encodes the combinatorics of the coarse circle packing for D. It also produces a list of faces of $\mathcal{K}$ enclosing a fundamental domain and it specifies the requisite geometry, in this case, hyperbolic.

4. `CirclePack` reads the combinatorics $\mathcal{K}$ and computes the associated hyperbolic packing label. This is a crucial step: theory gives the surface s_D. In practice, this is represented as a unique "packing label" for $\mathcal{K}$ consisting of the hyperbolic radii for $\mathcal{P}_\mathcal{K}$. `CirclePack` computes the packing label using a modification of an iterative algorithm suggested by Thurston, an algorithm which has been proven to converge [4].

5. Using the hyperbolic radii, combinatorics, and red chain of the packing, `CirclePack` will lay out a configuration $\mathcal{P}$ of circles in the hyperbolic plane whose carrier is the desired fundamental domain of the covering of s_D, as in Figure 5.11(a).

6. `DesPack` and `CirclePack` can communicate to permit display of various information in the hyperbolic plane: e.g., drawing the dessin, shading faces, labeling the 0- 1- and ∞-points, etc.

7. `CirclePack` will generate the combinatorics for successive hexagonal refinements $\mathcal{K}_n$, compute their packing labels, lay out their fundamental domains, and in conjunction with `DesPack`, carry out various drawing operations, as in Figure 5.11(b).

The difficulty in building examples typically lies in specifying dessin combinatorics and a red chain for visual layout. With this in hand, the remaining operations are quite fast. The sequence of commands to `CirclePack` is typically entered in a "script" file, which can be shared with others, easily modified, and invoked automatically.

10.2. The algorithm

The most computationally intense stage in working with discrete conformal structures involves computation of packing radii for circle packing complexes $\mathcal{K}$. The "packing" algorithm in `CirclePack` is iterative, similar in spirit to classical ralaxation methods for solving the discrete Laplace equation: the routine repeatedly passes through the list of circles, adjusting the radius of each circle it visits to make it fit with its immediate neighbors. The algorithm is described briefly in [**20**, p. 316]; for additional details and improvements in implementation due to Chuck Collins and the second author, see [**15**].

Tables 1, 2, and 3 list some details on our dessin examples. "Iterations" refers to number of passes through the full set of vertices during the iterative packing computation, so its product with the previous column indicates the total number of radii adjustments. Our runs were done on vanilla desktop PC's running linux. Packing times for the coarse dessins were essentially negligible, none required more than one second of cpu time (disregarding the overhead of input/output), and although the computation times increase rapidly with the number of circles and the desired numerical accuracy, even the most refined stages of our examples required no more than 10 minutes. Packings with many thousands of circles and complicated combinatorics can be more time consuming. One faces more delicate numerical issues in these large packings (see [**15**]), but `CirclePack` handles most without difficulty.

The examples in the table represent roughly 9 digit accuracy. To suggest the effect of desired accuracy, Table 3 lists relative timings for the stage-3 Klein surface (64,508 circles) for tolerances set from 10^{-5} down to 10^{-11}.

The properties of packing labels and the behavior of the packing algorithm are quite fascinating in their own right, and significant improvements may be possible with deeper understanding. For instance, radii adjustments may be modeled as discrete Markov processes having to do with the flow of "curvature" among vertices in the packing complex; see [**47, 48, 14**]. Packings of refinements, important if one wishes to approximate the classical objects, may be improved through multi-grid methods in which the labels for coarser packings provide the initial guesses for labels of their refinements; see [**20**, §5]. Finally, the algorithms are amenable to significant parallelization.

10.3. Accuracy

In Chapter 6 we gave some numerical results from our examples to suggest the accuracy of the discrete methods. Recall that this involved both computational and theoretical accuracy issues.

As for computational accuracy, consider a given circle packing complex $\mathcal{K}$ triangulating a compact surface. The associated packing label is effectively computable, meaning that in theory the packing algorithm can produce an approximation to the packing label having any desired level of accuracy. In addition, geometric considerations and certain monotonicity properties associated with circle configurations provide upper estimates on the errors.

In practice, the accuracy of computed packing labels is inferred from the accuracy of "angle sums". The angle sum at a vertex v is the sum of the angles in the faces to which v belongs; each of these angles is computed from the law of cosines using the labels for v and the other two vertices. It measures how closely the immediate neighbors come to wrapping round v.

Inaccuracies also arises when the circle packing (or its fundamental domain) is actually laid out on the plane, the disc, or sphere (which is normally projected from the disc). Due to inevitable errors in computed radii, there is in fact no consistent way to lay down an associated configuration of circles meeting the requisite tangency pattern. In practice, however, the circle centers are computed sequentially in successive generations from an initial circle; errors frequently tend to cancel one another, and an essentially consistent packing generally emerges. Fissures which can occur in the pattern as artifacts of the plotting sequence can generally be removed by reducing the error tolerance in the packing routine and by employing various averaging techniques to evenly distribute the roundoff errors during layout. (The tables above reflect 9 decimal place accuracy, though 5 places will generally yield a coherent packing.)

Example # Genus	Refinement Stage	Circle count	Iterations
Ex.#2, $g=0$ (lollypop)	coarse	25	31
	1	97	66
	2	385	319
	3	1537	1325
	4	6145	5017
Ex.#3, $g=0$ (one-arm)	coarse	74	27
	1	290	68
	2	1154	251
	3	4610	1234
	4	18434	5019
Ex.#4, $g=0$ (tree)	coarse	109	11
	1	433	36
	2	1729	104
	3	6913	524
	4	27650	2158
Ex.#5, $g=1$ ($\sqrt{7}$)	coarse	144	41
	1	576	94
	2	2304	213
	3	9216	580
	4	36864	1656
Ex.#6, $g=1$ ($\sqrt{7}$-conj.)	coarse	144	55
	1	576	171
	2	2304	389
	3	9216	1083
	4	36864	3374

TABLE 1. Computation details, genus 0 and 1

Example # Genus	Refinement Stage	Circle count	Iterations
Ex.#7, $g=2$ (peace)	coarse	82	36
	1	334	98
	2	1342	325
	3	5374	1273
	4	21502	4004
Ex.#8, $g=2$ (peace 2)	coarse	82	36
	1	334	108
	2	1342	346
	3	5374	1100
	4	21502	4568
Ex.#9, $g=2$ (generic)	coarse	250	96
	1	1006	332
	2	4030	1146
	3	16126	4234
Ex.#10, $g=3$ (Klein)	coarse	1004	110
	1	4028	406
	2	16124	1544
	3	64508	4776
Ex.#11, $g=3$ (Picard: Quine)	coarse	284	49
	1	1148	146
	2	4604	494
	3	18428	1724
	4	73724	6049
Ex.#11, $g=3$ (Picard: Sabbat & Voevodsky)	coarse	140	34
	1	572	73
	2	2300	261
	3	9212	1048
	4	36860	3486
Ex.#12, $g=4$ (generic)	coarse	522	94
	1	2106	248
	2	8442	952
	3	33786	3238

TABLE 2. Computation details, higher genus

Tolerance	10^{-5}	10^{-7}	10^{-9}	10^{-11}
Iterations	1564	2998	4662	6382
CPU time (mm:ss)	1:55	5:41	6:02	7:16

TABLE 3. Timings for the stage-3 Klein surface.

Bibliography

[1] Tom M. Apostol, *Modular functions and Dirichlet series in number theory*, 2nd ed., Graduate Texts in Mathematics, vol. 41, Springer Verlag, New York, 1990.
[2] Michel Bauer and Claude Itzykson, *Triangulations*, The Grothendieck Theory of Dessins d'Enfants (Leila Schneps, ed.), London Math. Soc. Lecture Note Series, vol. 200, Cambridge Univ. Press, Cambridge, 1994, pp. 179–236.
[3] A. F. Beardon, *A primer on Riemann surfaces*, London Math. Soc. Lecture Note Series, vol. 78, Cambridge Univ. Press, Cambridge, 1984.
[4] Alan F. Beardon and Kenneth Stephenson, *The uniformization theorem for circle packings*, Indiana Univ. Math. J. **39** (1990), 1383–1425.
[5] Lipman Bers, *Finite dimensional Teichmüller spaces and generalizations*, Bull. AMS **5** (1981), 131–172.
[6] Philip L. Bowers and Kenneth Stephenson, *Circle packings in surfaces of finite type: An in situ approach with applications to moduli*, Topology **32** (1993), 157–183.
[7] _____, *A branched Andreev-Thurston theorem for circle packings of the sphere*, Proc. London Math. Soc. (3) **73** (1996), 185–215.
[8] _____, *A "regular" pentagonal tiling of the plane*, Conformal Geometry and Dynamics **1** (1997), 58–86.
[9] James W. Cannon, *The theory of negatively curved spaces and groups*, Ergodic Theory, Symbolic Dynamics and Hyperbolic Spaces (Tim Bedford, Michael Keane, and Caroline Series, eds.), Oxford Science Publications, Oxford-NewYork-Tokyo, 1991, Comments on circle packing, pp. 315–369.
[10] _____, *The combinatorial Riemann mapping theorem*, Acta Mathematica **173** (1994), 155–234.
[11] James W. Cannon, W. J. Floyd, and Walter Parry, *Sufficiently rich families of planar rings*, Ann. Acad. Sci. Fenn. **24** (1999), 256–304, preprint, 1996.
[12] James W. Cannon, W. J. Floyd, and Walter Parry, *Finite subdivision rules*, Conformal Geometry and Dynamics **5** (2001), 153–196.
[13] James W. Cannon, William J. Floyd, Richard Kenyon, and Walter R. Parry, *Constructing rational maps from subdivision rules*, preprint, 2001.
[14] Charles R. Collins, Tobin A. Driscoll, and Kenneth Stephenson, *Curvature flow in conformal mapping*, preprint, 2003.
[15] Chuck Collins and Kenneth Stephenson, *A circle packing algorithm*, Computational Geometry: Theory and Applications **25** (2003), 233–256.
[16] Jean-Marc Couveignes and Louis Granboulan, *Dessins from a geometric point of view*, The Grothendieck Theory of Dessins d'Enfants (Leila Schneps, ed.), London Math. Soc. Lecture Note Series, vol. 200, Cambridge Univ. Press, Cambridge, 1994, pp. 79–113.
[17] H. S. M. Coxeter, *Regular complex polytopes*, Cambridge Univ. Press, Cambridge, New York, 1991.
[18] A. Douady and John Hubbard, *A proof of thurston's topological characterization of rational functions*, Acta. Math. **171** (1993), 263–297.
[19] Tomasz Dubejko, *Recurrent random walks, Liouville's theorem, and circle packings*, Math. Proc. Cambridge Philos. Soc. **121** (1997), no. 3, 531–546.
[20] Tomasz Dubejko and Kenneth Stephenson, *Circle packing: Experiments in discrete analytic function theory*, Experimental Mathematics **4** (1995), no. 4, 307–348.
[21] Hershel M. Farkas and Irwin Kra, *Riemann surfaces*, Graduate Texts in Math, vol. 71, Springer-Verlag, New York, 1980.

[22] Robert Fricke and Felix Klein, *Vorlesungen über die Theorie der automorphen Functionen*, vol. I and II, B. G. Teubner, 1897-1912.
[23] A. Grothendieck, *Esquisse d'un programme*, (1985), Preprint: introduced Dessins.
[24] Zheng-Xu He, *Rigidity of infinite disk patterns*, Ann. of Math. **149** (1999), 1–33.
[25] Zheng-Xu He and Oded Schramm, *On the convergence of circle packings to the Riemann map*, Invent. Math. **125** (1996), 285–305.
[26] Craig D. Hodgson and Igor Rivin, *A characterization of compact convex polyhedra in hyperbolic 3-space*, Invent. Math. **111** (1993), 77–111.
[27] G. Jones and D. Singerman, *Complex functions: An algebraic and geometric viewpoint*, Cambridge Univ. Press, Cambridge, New York, 1987.
[28] G. A. Jones and M. Streit, *Galois groups, monodromy groups and cartographic groups*, preprint.
[29] Gareth A. Jones, *Graph embeddings and maps on surfaces 1*, Mathematica Slovaca **47** (1997), no. 1, 1–33.
[30] F. Klein and R. Fricke, *Vorlesungen über die Theorie der elliptischen Modulfunctionen*, vol. 1, 2, Tuebner, Leipzig, 1890.
[31] G. Königs, *Recherches sur les intégrales de certaines èquationes functionelles*, Annales Ecole Normale Superior (3) **1** (1884), 3–41, Supplément.
[32] S. Kravetz, *On the geometry of Teichmüller spaces and the structure of their modular groups*, Ann. Acad. Sci. Fenn. **A1278** (1959), 1–35.
[33] O. Lehto and K. I. Virtanen, *Quasiconformal mappings in the plane*, 2nd ed., Springer-Verlag, New York, 1973.
[34] Olli Lehto, *Univalent functions and Teichmuller spaces*, Springer-Verlag, New York, 1987.
[35] M. Linch, *A comparison of metrics on Teichmüller space*, Proc. AMS **43** (1974), 349–352.
[36] Wilhelm Magnus, *Noneuclidean tesselations and their groups*, Academic Press, New York, London, 1974.
[37] Howard Masur, *On a class of geodesics in Teichmüller space*, Annals of Math. **102** (1975), no. 2, 205–221.
[38] G. J. McCaughan, *Some topics in circle packing*, Ph.D. thesis, University of Cambridge (advisor: Keith Carne), 1996.
[39] John W. Morgan, *On Thurston's uniformization theorem for three-dimensional manifolds*, The Smith Conjecture (Hyman Bass and John W. Morgan, eds.), Academic Press, New York, 1984, pp. 37–126.
[40] Subhashis Nag, *The complex analytic theory of Teichmüller spaces*, Canadian Mathematical Society Series of Monographs and Advanced Texts, Wiley-Interscience, John Wiley and Sons, New York, 1988.
[41] J. R. Quine, *Jacobian of the Picard curve*, Extremal Riemann surfaces, Contemporary Math, vol. 201, Amer. Math. Soc., Providence, 1997, pp. 33–41.
[42] John G. Ratcliffe, *Foundations of hyperbolic manifolds*, Graduate Texts in Math, vol. 149, Springer-Verlag, New York, Heidelberg, Berlin, 1994.
[43] Burt Rodin and Dennis Sullivan, *The convergence of circle packings to the Riemann mapping*, J. Differential Geometry **26** (1987), 349–360.
[44] L. Schneps (ed.), *The Grothendieck theory of dessins d'enfants*, London Math. Soc. Lecture Note Series, vol. 200, Cambridge Univ. Press, Cambridge, 1994.
[45] G. B. Shabat and V. A. Voevodsky, *Drawing curves over number fields*, The Grothendieck Festschrift, Vol. III (Boston, MA), Birkhauser Boston, 1990, pp. 199–227.
[46] Joel H. Shapiro, *Composition operators and classical function theory*, Universitext: Tracts in Mathematics, Sringer-Verlag, New York, Heidelberg, Berlin, 1993.
[47] Kenneth Stephenson, *Circle packings in the approximation of conformal mappings*, Bulletin, Amer. Math. Soc. (Research Announcements) **23, no. 2** (1990), 407–415.
[48] _____, *A probabilistic proof of Thurston's conjecture on circle packings*, Rendiconti del Seminario Mate. e Fisico di Milano **LXVI** (1996), 201–291.
[49] _____, *Approximation of conformal structures via circle packing*, Computational Methods and Function Theory 1997, Proceedings of the Third CMFT Conference (N. Papamichael, St. Ruscheweyh, and E. B. Saff, eds.), vol. 11, World Scientific, 1999, pp. 551–582.
[50] William Thurston, *The geometry and topology of 3-manifolds*, Princeton University Notes, preprint.

[51] _____, *The finite Riemann mapping theorem*, 1985, Invited talk, An International Symposium at Purdue University on the occasion of the proof of the Bieberbach conjecture, March 1985.

[52] Lloyd N. Trefethen and Tobin A. Driscoll, *Schwarz-Christoffel mapping in the computer era*, Proceedings of the International Congress of Mathematicians. Vol. III., Documenta Mathematica, Bielefeld, Bielefeld, 1998, Held in Berlin, August 1998, pp. 533–542.

[53] George Brock Williams, *Discrete approximation of conformal weldings using circle packings*, Indiana Univ. Math. J., to appear.

[54] _____, *Approximation of quasisymmetries using circle packings*, Discrete Comput. Geom. **25** (2001), 103–124.

[55] _____, *Earthquakes and circle packings*, J. Anal. Math. **85** (2001), 371–396.

[56] Joseph A. Wolf, *Spaces of constant curvature*, Publish or Perish Inc., Wilmington, DL, 1984.

[57] Jürgen Wolfart, *Mirror-invariant triangulations of Riemann surfaces, triangle groups, and Grothendieck dessins. variations on a theme of Belyi*, (1992), Frankfurt Preprint.

Editorial Information

To be published in the *Memoirs*, a paper must be correct, new, nontrivial, and significant. Further, it must be well written and of interest to a substantial number of mathematicians. Piecemeal results, such as an inconclusive step toward an unproved major theorem or a minor variation on a known result, are in general not acceptable for publication. Papers appearing in *Memoirs* are generally longer than those appearing in *Transactions*, which shares the same editorial committee.

As of March 1, 2004, the backlog for this journal was approximately 4 volumes. This estimate is the result of dividing the number of manuscripts for this journal in the Providence office that have not yet gone to the printer on the above date by the average number of monographs per volume over the previous twelve months, reduced by the number of volumes published in four months (the time necessary for preparing a volume for the printer). (There are 6 volumes per year, each containing at least 4 numbers.)

A Consent to Publish and Copyright Agreement is required before a paper will be published in the *Memoirs*. After a paper is accepted for publication, the Providence office will send a Consent to Publish and Copyright Agreement to all authors of the paper. By submitting a paper to the *Memoirs*, authors certify that the results have not been submitted to nor are they under consideration for publication by another journal, conference proceedings, or similar publication.

Information for Authors

Memoirs are printed from camera copy fully prepared by the author. This means that the finished book will look exactly like the copy submitted.

The paper must contain a *descriptive title* and an *abstract* that summarizes the article in language suitable for workers in the general field (algebra, analysis, etc.). The *descriptive title* should be short, but informative; useless or vague phrases such as "some remarks about" or "concerning" should be avoided. The *abstract* should be at least one complete sentence, and at most 300 words. Included with the footnotes to the paper should be the 2000 *Mathematics Subject Classification* representing the primary and secondary subjects of the article. The classifications are accessible from www.ams.org/msc/. The list of classifications is also available in print starting with the 1999 annual index of *Mathematical Reviews*. The Mathematics Subject Classification footnote may be followed by a list of *key words and phrases* describing the subject matter of the article and taken from it. Journal abbreviations used in bibliographies are listed in the latest *Mathematical Reviews* annual index. The series abbreviations are also accessible from www.ams.org/publications/. To help in preparing and verifying references, the AMS offers MR Lookup, a Reference Tool for Linking, at www.ams.org/mrlookup/. When the manuscript is submitted, authors should supply the editor with electronic addresses if available. These will be printed after the postal address at the end of the article.

Electronically prepared manuscripts. The AMS encourages electronically prepared manuscripts, with a strong preference for $\mathcal{AMS}$-LaTeX. To this end, the Society has prepared $\mathcal{AMS}$-LaTeX author packages for each AMS publication. Author packages include instructions for preparing electronic manuscripts, the *AMS Author Handbook*, samples, and a style file that generates the particular design specifications of that publication series. Though $\mathcal{AMS}$-LaTeX is the highly preferred format of TeX, author packages are also available in $\mathcal{AMS}$-TeX.

Authors may retrieve an author package from e-MATH starting from www.ams.org/tex/ or via FTP to ftp.ams.org (login as anonymous, enter username as password, and type cd pub/author-info). The *AMS Author Handbook* and the *Instruction Manual* are available in PDF format following the author packages link from www.ams.org/tex/. The author package can be obtained free of charge by sending email to pub@ams.org (Internet) or from the Publication Division, American Mathematical Society, 201 Charles St., Providence, RI 02904, USA. When requesting an author package, please specify $\mathcal{AMS}$-LaTeX or $\mathcal{AMS}$-TeX, Macintosh or IBM (3.5) format, and the publication in which your paper will appear. Please be sure to include your complete mailing address.

Sending electronic files. After acceptance, the source file(s) should be sent to the Providence office (this includes any TeX source file, any graphics files, and the DVI or PostScript file).

Before sending the source file, be sure you have proofread your paper carefully. The files you send must be the EXACT files used to generate the proof copy that was accepted for publication. For all publications, authors are required to send a printed copy of their paper, which exactly matches the copy approved for publication, along with any graphics that will appear in the paper.

TeX files may be submitted by email, FTP, or on diskette. The DVI file(s) and PostScript files should be submitted only by FTP or on diskette unless they are encoded properly to submit through email. (DVI files are binary and PostScript files tend to be very large.)

Electronically prepared manuscripts can be sent via email to pub-submit@ams.org (Internet). The subject line of the message should include the publication code to identify it as a Memoir. TeX source files, DVI files, and PostScript files can be transferred over the Internet by FTP to the Internet node e-math.ams.org (130.44.1.100).

Electronic graphics. Comprehensive instructions on preparing graphics are available at www.ams.org/jourhtml/graphics.html. A few of the major requirements are given here.

Submit files for graphics as EPS (Encapsulated PostScript) files. This includes graphics originated via a graphics application as well as scanned photographs or other computer-generated images. If this is not possible, TIFF files are acceptable as long as they can be opened in Adobe Photoshop or Illustrator. No matter what method was used to produce the graphic, it is necessary to provide a paper copy to the AMS.

Authors using graphics packages for the creation of electronic art should also avoid the use of any lines thinner than 0.5 points in width. Many graphics packages allow the user to specify a "hairline" for a very thin line. Hairlines often look acceptable when proofed on a typical laser printer. However, when produced on a high-resolution laser imagesetter, hairlines become nearly invisible and will be lost entirely in the final printing process.

Screens should be set to values between 15% and 85%. Screens which fall outside of this range are too light or too dark to print correctly. Variations of screens within a graphic should be no less than 10%.

Inquiries. Any inquiries concerning a paper that has been accepted for publication should be sent directly to the Electronic Prepress Department, American Mathematical Society, 201 Charles St., Providence, RI 02904, USA.

Editors

This journal is designed particularly for long research papers, normally at least 80 pages in length, and groups of cognate papers in pure and applied mathematics. Papers intended for publication in the *Memoirs* should be addressed to one of the following editors. In principle the Memoirs welcomes electronic submissions, and some of the editors, those whose names appear below with an asterisk (*), have indicated that they prefer them. However, editors reserve the right to request hard copies after papers have been submitted electronically. Authors are advised to make preliminary email inquiries to editors about whether they are likely to be able to handle submissions in a particular electronic form.

*Algebra to ROBERT GURALNICK, Department of Mathematics, University of Southern California, Los Angeles, CA 90089-1113; email: guralnic@math.usc.edu

Algebraic geometry to DAN ABRAMOVICH, Department of Mathematics, Boston University, 111 Cummington St., Boston, MA 02215; email: abramovic@bu.edu

*Algebraic number theory to V. KUMAR MURTY, Department of Mathematics, University of Toronto, 100 St. George Street, Toronto, ON M5S 1A1, Canada; email: murty@math.toronto.edu

Combinatorics and Lie theory to SERGEY FOMIN, Department of Mathematics, University of Michigan, Ann Arbor, Michigan 48109-1109; email: fomin@umich.edu

Complex analysis and complex geometry to DUONG H. PHONG, Department of Mathematics, Columbia University, 2990 Broadway, New York, NY 10027-0029; email: phong@math.columbia.edu

*Differential geometry and global analysis to LISA C. JEFFREY, Department of Mathematics, University of Toronto, 100 St. George St., Toronto, ON Canada M5S 3G3; email: jeffrey@math.toronto.edu

Dynamical systems and ergodic theory to ROBERT F. WILLIAMS, Department of Mathematics, University of Texas, Austin, Texas 78712-1082; email: bob@math.utexas.edu

*Functional analysis and operator algebras to MARIUS DADARLAT, Department of Mathematics, Purdue University, 150 N. University St., West Lafayette, IN 47907-2067; email: mdd@math.purdue.edu

*Geometric analysis to TOBIAS COLDING, Courant Institute, New York University, 251 Mercer St., New York, NY 10012; email: colding@cims.nyu.edu

*Geometric analysis to MLADEN BESTVINA, Department of Mathematics, University of Utah, 155 South 1400 East, JWB 233, Salt Lake City, Utah 84112-0090; email: bestvina@math.utah.edu

Harmonic analysis to ALEXANDER NAGEL, Department of Mathematics, University of Wisconsin, 480 Lincoln Drive, Madison, WI 53706-1313; email: nagel@math.wisc.edu

Harmonic analysis, representation theory, and Lie theory to ROBERT J. STANTON, Department of Mathematics, The Ohio State University, 231 West 18th Avenue, Columbus, OH 43210-1174; email: stanton@math.ohio-state.edu

*Logic to STEFFEN LEMPP, Department of Mathematics, University of Wisconsin, 480 Lincoln Drive, Madison, Wisconsin 53706-1388; email: lempp@math.wisc.edu

Number theory to HAROLD G. DIAMOND, Department of Mathematics, University of Illinois, 1409 W. Green St., Urbana, IL 61801-2917; email: diamond@math.uiuc.edu

*Ordinary differential equations, and applied mathematics to PETER W. BATES, Department of Mathematics, Michigan State University, East Lansing, MI 48824-1027; email: peter@math.msu.edu

*Partial differential equations to PATRICIA E. BAUMAN, Department of Mathematics, Purdue University, West Lafayette, IN 47907-1395; email: bauman@math.purdue.edu

*Probability and statistics to KRZYSZTOF BURDZY, Department of Mathematics, University of Washington, Box 354350, Seattle, Washington 98195-4350; email: burdzy@math.washington.edu

*Real analysis and partial differential equations to DANIEL TATARU, Department of Mathematics, University of California, Berkeley, Berkeley, CA 94720; email: tataru@ math.berkeley.edu

All other communications to the editors should be addressed to the Managing Editor, WILLIAM BECKNER, Department of Mathematics, University of Texas, Austin, TX 78712-1082; email: beckner@math.utexas.edu.

Titles in This Series

807 **Carlos A. Cabrelli, Christopher Heil, and Ursula M. Molter,** Self-similarity and multiwavelets in higher dimensions, 2004

806 **Spiros A. Argyros and Andreas Tolias,** Methods in the theory of hereditarily indecomposable Banach spaces, 2004

805 **Philip L. Bowers and Kenneth Stephenson,** Uniformizing dessins and Belyĭ maps via circle packing, 2004

804 **A. Yu. Ol'shanskii and M. V. Sapir,** The conjugacy problem and Higman embeddings, 2004

803 **Michael Field and Matthew Nicol,** Ergodic theory of equivariant diffeomorphisms: Markov partitions and stable ergodicity, 2004

802 **Martin W. Liebeck and Gary M. Seitz,** The maximal subgroups of positive dimension in exceptional algebraic groups, 2004

801 **Fabio Ancona and Andrea Marson,** Well-posedness for general 2×2 systems of conservation laws, 2004

800 **V. Poénaru and C. Tanasi,** Equivariant, almost-arborescent representations of open simply-connected 3-manifolds; A finiteness result, 2004

799 **Barry Mazur and Karl Rubin,** Kolyvagin systems, 2004

798 **Benoît Mselati,** Classification and probabilistic representation of the positive solutions of a semilinear elliptic equation, 2004

797 **Ola Bratteli, Palle E. T. Jorgensen, and Vasyl' Ostrovs'kyĭ,** Representation theory and numerical AF-invariants, 2004

796 **Marc A. Rieffel,** Gromov-Hausdorff distance for quantum metric spaces/Matrix algebras converge to the sphere for quantum Gromov-Hausdorff distance, 2004

795 **Adam Nyman,** Points on quantum projectivizations, 2004

794 **Kevin K. Ferland and L. Gaunce Lewis, Jr.,** The $RO(G)$-graded equivariant ordinary homology of G-cell complexes with even-dimensional cells for $G = \mathbb{Z}/p$, 2004

793 **Jindřich Zapletal,** Descriptive set theory and definable forcing, 2004

792 **Inmaculada Baldomá and Ernest Fontich,** Exponentially small splitting of invariant manifolds of parabolic points, 2004

791 **Eva A. Gallardo-Gutiérrez and Alfonso Montes-Rodríguez,** The role of the spectrum in the cyclic behavior of composition operators, 2004

790 **Thierry Lévy,** Yang-Mills measure on compact surfaces, 2003

789 **Helge Glöckner,** Positive definite functions on infinite-dimensional convex cones, 2003

788 **Robert Denk, Matthias Hieber, and Jan Prüss,** $\mathcal{R}$-boundedness, Fourier multipliers and problems of elliptic and parabolic type, 2003

787 **Michael Cwikel, Per G. Nilsson, and Gideon Schechtman,** Interpolation of weighted Banach lattices/A characterization of relatively decomposable Banach lattices, 2003

786 **Arnd Scheel,** Radially symmetric patterns of reaction-diffusion systems, 2003

785 **R. R. Bruner and J. P. C. Greenlees,** The connective K-theory of finite groups, 2003

784 **Desmond Sheiham,** Invariants of boundary link cobordism, 2003

783 **Ethan Akin, Mike Hurley, and Judy A. Kennedy,** Dynamics of topologically generic homeomorphisms, 2003

782 **Masaaki Furusawa and Joseph A. Shalika,** On central critical values of the degree four L-functions for GSp(4): The Fundamental Lemma, 2003

781 **Marcin Bownik,** Anisotropic Hardy spaces and wavelets, 2003

780 **S. Marmi and D. Sauzin,** Quasianalytic monogenic solutions of a cohomological equation, 2003

779 **Hansjörg Geiges,** h-principles and flexibility in geometry, 2003

TITLES IN THIS SERIES

778 **David B. Massey,** Numerical control over complex analytic singularities, 2003
777 **Robert Lauter,** Pseudodifferential analysis on conformally compact spaces, 2003
776 **U. Haagerup, H. P. Rosenthal, and F. A. Sukochev,** Banach embedding properties of non-commutative L^p-spaces, 2003
775 **P. Lochak, J.-P. Marco, and D. Sauzin,** On the splitting of invariant manifolds in multidimensional near-integrable Hamiltonian systems, 2003
774 **Kai A. Behrend,** Derived ℓ-adic categories for algebraic stacks, 2003
773 **Robert M. Guralnick, Peter Müller, and Jan Saxl,** The rational function analogue of a question of Schur and exceptionality of permutation representations, 2003
772 **Katrina Barron,** The moduli space of $N=1$ superspheres with tubes and the sewing operation, 2003
771 **Shigenori Matsumoto,** Affine flows on 3-manifolds, 2003
770 **W. N. Everitt and L. Markus,** Elliptic partial differential operators and symplectic algebra, 2003
769 **Jie Wu,** Homotopy theory of the suspensions of the projective plane, 2003
768 **R. Höpfner and E. Löcherbach,** Limit theorems for null recurrent Markov processes, 2003
767 **Po Hu,** S-modules in the category of schemes, 2003
766 **Su Gao and Alexander S. Kechris,** On the classification of Polish metric spaces up to isometry, 2003
765 **Robert Bieri and Ross Geoghegan,** Connectivity properties of group actions on non-positively curved spaces, 2003
764 **J. Spandaw,** Noether-Lefschetz problems for degeneracy loci, 2003
763 **Yasuyuki Kachi and Eiichi Sato,** Segre's reflexivity and an inductive characterization os hyperquadrics, 2002
762 **Leiba Rodman, Ilya M. Spitkovsky, and Hugo Woerdeman,** Abstract band method via factorization, positive and band extensions of multivariable almost periodic matrix functions, and spectral estimation, 2002
761 **Oliver Druet and Emmanuel Hebey,** The AB program in geometric analysis : Sharp Sobolev inequalities and related problems, 2002
760 **Markus Banagl,** Extending intersection homology type invarients to non-Witt spaces, 2002
759 **Donald M. Davis,** From representation theory to homotopy groups, 2002
758 **Alan Forrest, John Hunton, and Johannes Kellendonk,** Topological invariants for projection method patterns, 2002
757 **Douglas Bowman,** q-difference operators, orthogonal polynomials, and symmetric expansions, 2002
756 **José Ignacio Cogolludo-Agustín,** Topological invariants of the complement to arrangements of rational plane curves, 2002
755 **M. A. Mandell and J. P. May,** Equivariant orthogonal spectra and S-modules, 2002
754 **Edward L. Green, Idun Reiten, and Øyvind Solberg,** Dualities on generalized Koszul algebras, 2002
753 **Daniel Panazzolo,** Desingularization of nilpotent singularities in families of planar vector fields, 2002
752 **Linus Kramer,** Homogeneous spaces, Tits buildings, and isoparametric hypersurfaces, 2002

For a complete list of titles in this series, visit the
AMS Bookstore at **www.ams.org/bookstore/**.